SE EY RAILWAY

An Illustrated History

MICHAEL LANCASHIRE

First published by Sannox Press in 2020

ISBN 978-0-9928130-8-6

Cover design by Michael Lancashire. Image used on front cover by Neil Scott; on back cover by Keith Wilkinson. Both released under Creative Commons, full details listed at Image Credits.

In memory of Brian Lancashire, steam enthusiast
6th April 1939 – 30th September 2019

Brian's fascination with steam began when he was a young boy and he would stop on his walk to school (for long enough to get in trouble on occasion!), lean over the side of Bromford Bridge and stare transfixed as the goods trains ran below him in and out of Birmingham.

When he was slightly older he was a founding member of the Bromford Locomotive Society where he made friends for life, served as treasurer, and travelled the length and breadth of Britain gaining an encyclopaedic knowledge not just of railways but of the geography of the country.

After Beeching took an axe to his precious steam and he couldn't find enough in Britain to satisfy him, he travelled several times around Europe in search of new engines and a decent meal with ingredients he could identify.

Later, when he had a family and couldn't travel as much he found other ways to satisfy his passion. First spending hours in muddy fields, with his precious daughter standing by his side (with just slightly less patience), waiting for a mainline steam special to pass; and then later on the Severn Valley. It was there that he passed his love of steam on to the generations who came after him. Thanks to him the SVR will always have a place in our hearts.

As you read this book, think of him.

CONTENTS

INTRODUCTION

Since their introduction in the 1800s railways have always seemed to have a peculiarly strong place in men's hearts.

The passion that they inspire is perhaps best illustrated by a quote from Gerald Nabarro (a complicated man, who we will meet in Chapter Six):

> *"The building of the railways, which* ***remain the greatest physical achievement carried out by the human race*** *within a comparatively short space of time..."*

These are strong words but perhaps given that you have picked this book up, you are inclined to agree with them? I might be. But they are all the more astonishing when you realise that he wrote them in 1971, *just two years after the human race walked on the moon*, which took place only twelve years after we first reached space!

So, railways have always provoked a passionate response and the Severn Valley Railway is no different.

Whilst it is not really my goal to convince anyone the above quote is an accurate reflection of the various achievements of mankind, I do hope that by the time you have read this book you will have caught some of the passion that makes someone say it.

The Severn Valley dates from 1862 though, as we will see, at the time it stretched much further than the current preserved line, running from Hartlebury to Shrewsbury.

Although the main focus of this book is the preservation era, in contrast to many histories of the line I have attempted to provide an overview from the very earliest days right up to the point of writing. To that end, the book is structured as follows:

Chapters One to Four cover the period from the idea of the line being conceived, through to the closures of the 1960s. We dwell a little on the fascinating men who built the line and include some diversions away from the Severn Valley Railway.

Clearly the world wars formed a central part of the history of twentieth century Britain, but what may be less well appreciated today is the scale of the involvement of the railways. Previous histories of the Severn Valley Railway have not touched on their impact, presumably because the SVR was not particularly involved directly. However, I believe if we are to understand the closures that followed, then it is vital that we know the context in which it happened. We therefore cover the railways' role in war, mobilising and supporting the war effort, in more detail than I originally expected when setting out to write this book.

Those who want to focus exclusively on the SVR line itself can skip Chapter Three and perhaps skim read Chapter Four (but as I warn you again in the opening of the chapter, you'll miss some fascinating stories).

Chapters Five to Seven cover the early history of the preservation era, telling the dramatic story of its rescue from neglect and bring the story up to 1984 when the line reached Kidderminster.

Along the way we'll cover a power struggle over control of the railway, which is belied by today's quiet, perhaps even quaint nature of the line.

Chapter Eight describes the stations along the line, giving a little of the history of each where it hasn't already been told in the broader story of the line. For those who like their history in bite size chunks, this might be a good place to start.

Chapter Nine brings the story up to date, covering a couple of key challenges the railway has faced since the line between Bridgnorth and Kidderminster was completed.

(And **Chapter Ten** is simply the brief thoughts of an author who felt he'd earned a chapter to sum up).

The book is organised broadly chronologically, but each chapter can stand alone, and references are provided throughout to point you to other relevant sections of the book should you desire.

One final point: there is a focus on steam rather than diesel throughout. This is for two reasons, firstly this is a history of the line and for most of its history steam was the main form of motive power; secondly, it is where my interest lies.

Where it is relevant to the greater arc of the story I have included references to the diesel fleet and infrastructure, but I fear I do not do it justice largely because I am not the best man for the job. A detailed history of the diesel preservation movement is definitely due, but it is beyond the scope of this book.

CHAPTER ONE: The Birth of the Severn Valley Railway

STEAM POWER

The connection of the Severn Valley area with steam locomotion goes right back to the very earliest days of steam power.

The first paying steam train, called the *Catch Me Who Can* (in light of its then incredible top speed of 12 miles per hour!) was designed by steam pioneer Richard Trevithick and built in 1808 by John Uperth Rastrick and John Hazledine in Hazeldine's foundry at Bridgnorth.

Trevithick ran the engine on a circular track built specially near Euston Square in London. His intention was to introduce the public to the idea of steam locomotion and to make money at the same time from the fare of 1 shilling per ride. Which of these goals was uppermost in his mind can probably be determined from the fact that he built a tall wooden fence around the track preventing anyone who hadn't paid from being able to catch sight of the engine.

As a money spinner it didn't work. The iron rails were too brittle to take the weight of the engine and they broke repeatedly in the couple of weeks the track was operational.

Trevithick, described by the Encyclopaedia Britannica as "entirely lacking in business sense" died penniless in 1833. Luckily for us his idea of steam powered locomotion did not die.

Four years before his death the Stephensons (father George and son Robert) had demonstrated their *Rocket* and set the world on track for the coming of the railways.

Advert for ***Catch Me Who Can****, the first paying steam engine, designed by Richard Trevithick and built in 1808 at Bridgnorth by John Rastrick and John Hazledine.*

Richard Trevithick *inventor and steam pioneer*

THE RAILWAY COMES TO THE SEVERN VALLEY

Railways in the nineteenth century, which was nearly all of the railways we have today, needed special acts of parliament to set them up.

Despite this there was no shortage of budding entrepreneurs looking to set them up and soon there was almost no part of the country which did not have its own railway.

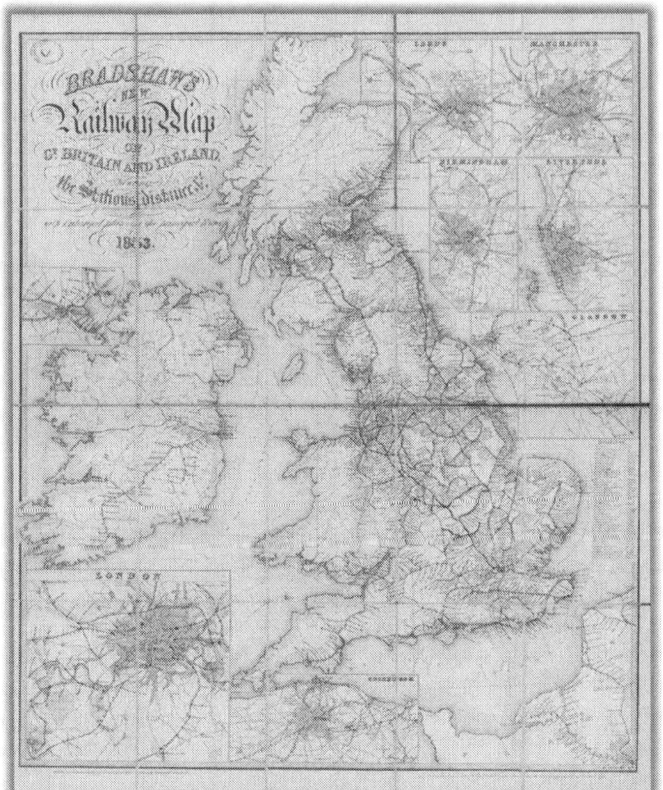

Bradshaw's Railway Map 1853, showing the extent to which Britain was in the grip of Railway Mania.

The idea of a line running roughly parallel to the River Severn dates from the late nineteenth century at a time known as Railway Mania.

Plans were drawn up on several occasions from 1801 onwards, including one in 1846 by Robert Stephenson himself (though as he was associated with a huge number of other railways at the time, we can be fairly sure he had little to do with it other than lending his name to the venture).

Most of these ideas came to nothing but, after several false starts, a Severn Valley Railway company was formed in 1852 which finally got the go ahead for a line in almost its current form in 1853.

The line that was agreed by Queen Victoria in 1853 ran from Shrewsbury to Hartlebury where it met up with the Oxford, Worcester and Wolverhampton line.

Of course, getting a bill through parliament and into law was only the first step. There was still the small matter of building the railway itself!

For that the directors had turned to a well-known railway man of the time, Sir Samuel Morton Peto.

A POWER STRUGGLE

Peto was a colourful Victorian entrepreneur who had made one of the largest fortunes of the time running a building company with his cousin Thomas Grissell. They were responsible for building the Houses of Parliament and Nelson's Column among many other famous landmarks.

***Sir Samuel Morton Peto** in middle age.*

A brief summary of Peto's achievements makes it obvious why the directors would have wanted him to support the line.

In 1846, six years before his involvement with the Severn Valley Railway, Peto and his cousin dissolved their construction company. After several successful railway contracts, including building the Wharncliffe Viaduct designed by Isambard Kingdom Brunel for the GWR, Peto's interests had shifted from building work to what he saw as the greater opportunities presented by the new-fangled railways. Grissell, despite their success with the GWR and other railway lines, didn't have Peto's taste for risk and preferred to continue working on safer, more traditional building contracts.

The houses of parliament which Peto's old firm built and he later worked in as an MP

After the split with his cousin, Peto formed a new company, Peto and Betts, with Edward Ladd Betts to concentrate on railway building. They carried out work on the Great Northern, the East Lincolnshire, and the Oxford, Worcester and Wolverhampton.

He served as an MP for three different constituencies between 1847 and 1868 (his interest in the Severn Valley began while he was still MP for Norwich and then overlapped with his longest break from parliament); he was a commissioner and guarantor of the Great Exhibition of 1851, designed to showcase British industrial ingenuity to the world; he was at various points during this period recognised as the largest employer in the world; and he was a truly international man, taking a key role in railways in Canada, Germany and Denmark among others.

In late 1854 and early 1855, with the war in the Crimea going badly for the British troops, Peto stepped in and organised the building of a military railway. He and his partners (the same Edward Betts and Thomas Brassey who were involved in the Severn Valley with him) did it all at cost and without requiring any contract from the government. In 1855 he was made a baronet in recognition of this.

Building the ***Grand Crimean Central Railway****. Navvies laid 29 miles in about 8 weeks. Peto's contribution to the Crimean War and the reason he was made a baronet.*

And rather surprisingly for a man who was so busy outside the home, he also managed to find the time to have fifteen children from his two marriages!

This then, was the man that the Severn Valley directors wanted to have on board.

Peto joins the Severn Valley

When they approached him to become involved Peto agreed and although he did not buy any shares himself during the initial offering, one of his long time partners, Thomas Brassey was the single largest subscriber, buying £30,000 worth.

Peto himself put forward £47,250 (or £4.7m in 2019 terms), more than half of the money needed to get the Bill considered by parliament.

Although nothing survives to confirm it, we can probably assume that the Severn Valley Railway directors' invitation to Peto to become the main contractor was driven by a desire to access his and his colleagues' capital and have his name associated with the line, thereby increasing the chances of it being approved.

Whatever the reason they invited him, some of them at least soon had reason to regret it.

Peto would no doubt be described today as an "activist investor", never one to sit quietly on the sidelines, he liked to take control of the companies he was connected to and the Severn Valley was no different.

In November 1853 he wrote to the Severn Valley Railway board and insisted that they spend as little money as possible until work began. (Work his company as the main contractor would eventually be paid for, we might note!) The chairman Jonathan Thorp agreed. In any case no work could yet begin because there were a number of changes needed to the planned course of the line which meant that they would have to go back to parliament to be approved in a new Bill.

The planning continued but by April 1854 matters came to a head.

There was a clause included in the new Bill which suggested that the current directors of the company should be automatically renewed until the line was finished. Unfortunately for Thorp, Peto got wind of this clause and was outraged at the removal of the shareholders power to vote for the board.

A brief power struggle ensued.

Peto wrote to Thorp on the 11th April telling him that he would oppose the new Bill in parliament and publicly call attention to the fact that the directors were seeking to undermine the shareholders' rights.

Opposition from a man of Peto's reputation in railways could have been fatal to the Bill and both men must have known it.

Thorp argued with Peto's solicitors for a fortnight but he could not stand against the man who by this time had become one of the largest shareholders of the company.

Finally on 3rd May Peto was made a director of the company and the offending Bill was withdrawn.

Two weeks later, on the 18th May, Thorp resigned and Peto was made the chairman of the board.

He remained as chairman until 1857. During his time in the chair he got two new Bills passed through parliament with no objections, largely due to the negotiating he did beforehand with all of the interested parties (even Thomas Whitmore, the owner of the Apley Park estate just north of Bridgnorth, who had been in a decade-long dispute with the railway, was satisfied!), however, the manner of his departure sheds some rather questionable light on his original motivations.

In July 1857 another engineering firm, Smith and Knight, offered to produce a quote to carry out the work on the Severn Valley. By now Peto had been assuming for years that his own company, working with Thomas Brassey, would be the main contractor. On the 28th July he once again sent a letter to the board. In it he reminded them that:

> *...I and my friends agreed to take Shares to the Amount of £225,000* ***upon the understanding that my friends were to be the Contractors*** *for the works...*

He also pointed out that he had become a director and then the Chairman in order to help the company out of financial difficulty and now that he had fixed those problems, he intended to:

> *"...**retain the Interest in the Company for which I have subscribed and to take the Contracts for the Works as originally intended**, provided the Company feel themselves in a situation to provide the additional funds requisite for carrying out the Undertaking...I think it is right now to retire from the Direction, and beg you will consider this letter as an intimation of my Resignation.*

That is, he'd keep his shares, resign from the board and build the railway, provided they could find enough money to pay him.

Peto's chairmanship should probably be considered successful, he did after all, secure the necessary funding to get the line built and shepherded the Bills through parliament to make sure that it happened. However, a cynical person with the benefit of historical perspective on Peto's other dealings might think that his main aim throughout had been to secure the work for his firm.

Peto, Brassey and Betts went on to build the Severn Valley line successfully under John Fowler as described below, but we cannot leave Peto without commenting on the ultimate end of this fascinating man's career.

Downfall

In line with Peto's appetite for high-risk and high-gain ventures Peto and Betts took on a number of railway contracts and agreed to operate the line themselves or asked to be paid in shares (including, interestingly to us, the SVR).

In 1863 they agreed to build the London, Chatham and Dover Railway on a share-only basis and Peto was appointed chief financial advisor to the railway.

The London, Chatham and Dover Railway (LCDR) insignia above Blackfriar's Bridge.

The LCDR was the immediate cause of Sir Samuel Morton Peto's fall from grace.

Some quite complicated financial shuffling now took place with loans backwards and forwards between Peto and Betts and the LCDR which ultimately meant the railway appeared to have a lot more money than it really did.

All probably would have been alright but, unfortunately for them, their main bank got into financial difficulties (which are interesting in their own right but don't concern us here) and was unable to make any payments and the whole house of cards came tumbling down.

The shares that Peto and Betts had been paid in became worthless, Peto was fired, his firm was declared bankrupt and the LCDR sued Peto and Betts as individuals for the sum of £6,661,941, 19 shillings and a penny, a mind blowing £600 million in 2019.

Despite his riches neither Peto's pocket nor his reputation could survive that sort of hit, he was forced to leave parliament and eventually even the country, emigrating for a time to Hungary. He died in 1889 in obscurity.

The mess that was left from the bankruptcy of Peto and Betts took thirty two years to resolve.

Sir Samuel Morton Peto
later in life.

JOHN FOWLER: BUILDING THE SVR

The fall of Peto was long in the future though and luckily had no impact on the Severn Valley Railway.

In 1855, following the untimely death of the original engineer, the directors hired another well respected railway man named John Fowler as chief engineer.

***John Fowler** chief engineer of the SVR (later Sir John Fowler, Baronet)*

Fowler was an equally impressive but much less flamboyant character than Peto, for a start he only managed to have four children!

Despite his lack of ambition in the home, his railway career was second to none. He was the engineer of Grosvenor Bridge at Pimlico, the first railway bridge across the Thames and the chief engineer of the first underground railway in the world, the Metropolitan Line in London.

Advert for the Metropolitan Line from the Illustrated London News, 27th December 1862

For his work on the underground he was paid approximately £300,000 (about £27.5m in today's money). He was Chief Engineer of the Oxford, Worcester and Wolverhampton that met the Severn Valley Railway at Hartlebury and then, while already engineer for the Severn Valley, he became chief engineer of GWR following Isambard Kingdom Brunel's death in 1859.

Above all of this though he is most famous for one bridge in particular (and oddly not the Victoria Bridge at Arley!) After the collapse of the Tay Bridge in 1879 Sir Thomas Boucher's design for a bridge across the firth of Forth was scrapped and replaced with Fowler's iconic design which is today a UNESCO World Heritage site and for which he received a baronetcy

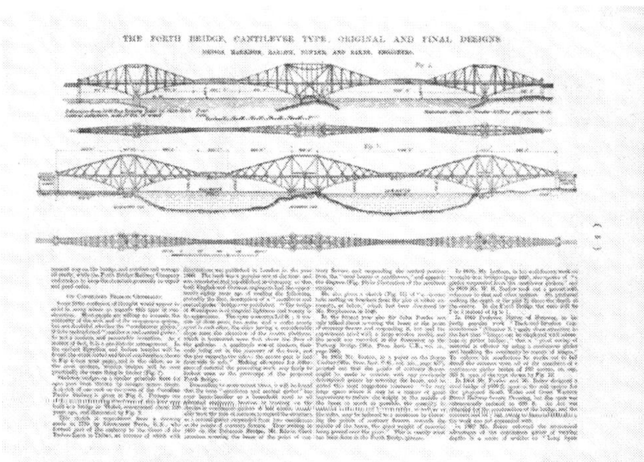

Design of the ***Forth Bridge*** *by John Fowler, as described in "ENGINEERING" February, 28th 1890*

Before he worked on his most famous project of all though Fowler oversaw the completion of the Severn Valley Railway.

Work began in August 1858 and went for the most part according to plan. There was a robbery, and an unfortunate death of one of the workers, but the thing that caused the greatest delay to the line was the condition of the ground.

The area has always been subject to landslips, indeed they have continued to plague the line up to the present day, but 1859 and 1860 were particularly wet years and there were a number of slips that held up the work. The biggest slip of all happened just north of Highley and meant that the course of the railway had to be changed. The sharp bends that can still be experienced as you leave Highley for Bridgnorth today are a direct result of this slip.

2-6-0 42968 showing off the ***sharp bends in the line caused by the landslips*** *as she approaches Highley from Bridgnorth*

The difficulties notwithstanding Fowler remained optimistic throughout. In August of 1861 he informed the directors that the line would be open by October of that year. Unfortunately when an inspection was carried out on behalf of the Board of Trade the inspector, a Colonel William Yolland, insisted on a number of changes before he could give permission for the line to open.

The next couple of months were spent addressing Yolland's objections, until finally he gave his approval of the changes and the formal opening took place on 31st January 1862, six years after Royal Assent had been given to the bill and almost two years after the October 1860 date that Fowler had originally suggested in 1859.

The Severn Valley Railway had arrived.

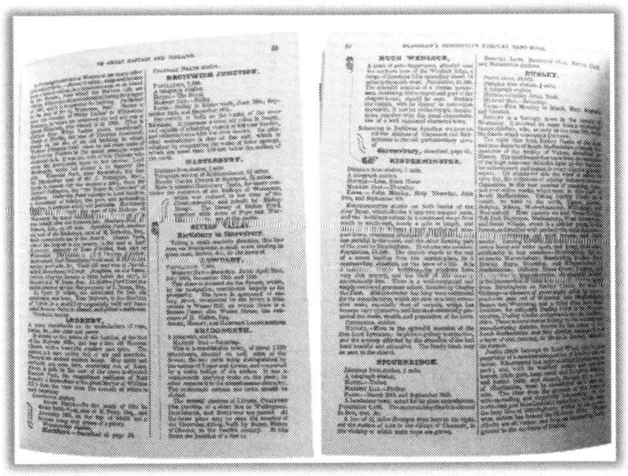

***The Severn Valley Railway formally opened on 31st January 1862**, the entry above shows the description in Bradshaw's Railway Guide the following year.*

CHAPTER TWO: The Railway until 1914

GREAT WESTERN RAILWAY

On 14th June 1860, whilst it was still under construction the SVR had been leased to the OW&W, and two days later the OW&W merged with two other railways to become the West Midland Railway so when the SVR started operating on 31st January 1862 it was as a part of the WMR.

A year and a half later, on 1st August 1863 the West Midland Railway was also dissolved and became a part of the Great Western Railway (GWR). GWR continued to operate the SVR as a separate railway until the 1st July 1871 when it formally became a branch line of the GWR and remained that way until nationalisation.

***Great Western Railway** coat of arms. The SVR became part of the GWR family in 1863 and in 1871 it ceased being a separate entity altogether and became a branch line of the GWR.*

The Great Western was a powerhouse of steam engine design and we will touch later on some of those engines, however, in terms of the line itself the most significant development under GWR ownership was the opening in 1878 of the Bewdley to Kidderminster Loop.

The SVR reaches Kidderminster

A line joining Bewdley to Kidderminster, known as the Kidderminster Loop, had first been authorised in a meeting on 6th February 1861, and was enshrined in law in the West Midland Severn Valley Companies' Act in August of that year, but neither the West Midland nor Great Western Railway were keen to embark on building it.

GWR actually favoured an alternative route between Bewdley and Stourbridge and resisted building the Kidderminster Loop until parliament rejected the Stourbridge option in 1873 and ordered them to start work on the Loop immediately. In a surprisingly churlish response GWR chairman Daniel Gooch over the next couple of months said that they were, "obliged to proceed with your Kidderminster curve, and finish the obligation you took up on yourselves… and which you have been putting off…" and referred to it as a useless curve that they (the GWR shareholders) didn't want but were now forced to build. So not a fan then.

Building the Loop wasn't without difficulties either, although only three miles long it required a great deal of work to accomplish, with a tunnel, a viaduct and considerable cuttings being necessary to reach Kidderminster. Sadly three men were killed in the construction of Bewdley tunnel, but finally in March 1878 the work was ready for inspection. In what must have felt like the last straw the line failed its first inspection.

It seemed that just about everybody had grown tired of the project and, when it was eventually opened on 1st June 1878, there was no fanfare.

At this point the line that we recognise today as the preserved line was complete.

OPERATIONS UNTIL WORLD WAR I

When the Loop opened it increased the volume of traffic on the line significantly as trains would operate throughout the day around the circuit formed between Hartlebury, Bewdley and Kidderminster. Apart from that the Severn Valley line was never a particularly busy passenger route. When it opened, just four trains a day operated each way between Hartlebury and Shrewsbury, by 1900 it had increased to five per day during the working week, only to be cut back again after the second world war.

There was, however, a steady stream of goods trains. With one service scheduled from Shrewsbury daily, another leaving Kidderminster to head for the Wyre Forest and additional trains from 1925 onwards from Foley Park to service the sugar factory, Kidderminster station was seldom empty of goods.

0-6-2T 6695 pulls a goods train at the 2010 Spring Gala. Although the 5600 class did not work the Severn Valley, goods trains would have been a familiar sight on the line.

The other consequence of the opening of the Kidderminster Loop was the introduction of the first true signal boxes onto the Severn Valley line.

Despite John Fowler's intention, the Severn Valley is almost entirely single track so to avoid accidents it is essential that access by any train to each section of line is controlled completely.

Signalling in the early days of the railways had been through single needle telegraph in which the operator of one station sent a message ahead to the operator at the next station informing him (and it was always a 'him' at this time) of the approach of a train. Unfortunately there was no connection between this signal and the points that controlled the line so the whole process relied on the signalmen getting it right.

Although Bewdley had been a junction since 1864 with the Tenbury line, the opening of the Loop meant that it now also served as a junction for the original Severn Valley line to Hartlebury and the new loop to Kidderminster. This meant that tighter controls were necessary to reduce the chance of accidents. As a result two signal boxes, Bewdley North and Bewdley South, were commissioned on 24th October 1877 and built in time for the opening of the Loop.

2-8-0 2857 passing ***Bewdley South signal box*** *pulling a Christmas special.*

In addition to the signals themselves, further control was provided by the 'staff and ticket' system in which each section of line had a dedicated staff which would be handed to the engine driver. The driver would then hand it off when he reached the next station.

The use of the staff meant that it wasn't possible for engines to enter the same section of line in opposite directions and made lines much safer, however, it also meant that if two trains were heading in the same direction the second would have to wait for the staff to be returned (on foot or by horse) before entering the next section. Obviously that wasn't efficient and so the 'ticket' system was introduced.

If there was more than one train heading in the same direction, the driver of the first would be shown the staff and given a ticket in its place, the driver of the last train would be given the staff itself.

This combined 'staff and ticket' setup was much safer than previous systems but still allowed efficient running and was the process in place in 1878 when the Loop was opened. In fact permission to open the Loop was conditional on the GWR confirming to the Board of Trade that it would be operated under 'staff and ticket'.

(Since preservation the SVR has been operated mainly by using an electric token system which is described in the section on Bewdley in Chapter Eight.)

Despite the obvious advantages it still took a little over twenty years for signal boxes to be installed at the remaining stations on the line, and even then judging by a minor accident in 1898 it is clear that the interlocking of signals and points hadn't taken place.

ENGINES ON THE LINE

It is not possible, and wouldn't make very good reading, to provide a list of all of the engines that were used on the line during this period, however, it is interesting to touch on some of the classes that were used on GWR branch lines like the Severn Valley. (Particularly as it makes for a very different experience to the one enjoyed by someone who has only known the line in preservation.)

0-6-0 Dean Goods

280 of the 0-6-0 2301 class were built between 1883 and 1893.

They were designed by William Dean in 1883 as a tender engine for mixed traffic, and they could be seen all over the GWR branch line network throughout this period and indeed much later. The last one was not withdrawn until 1953, five years after nationalisation.

Despite the fact that they did work both passenger and freight services as Dean had intended, they became most strongly associated with freight, earning the nickname 'Dean Goods'.

3205, an 0-6-0 tender engine built in 1949 to ***a design which had evolved from the nineteenth century 2301 class, 'Dean Goods'.***

Sadly only one example of the original class can be seen today, and that not in steam. It is housed as a static display in the GWR Museum at Swindon. However, Dean Goods formed the basis of a number of later engine classes that worked the SVR line, including Hawksworth's 3200 class, one of which, 3205, pulled the first train on the SVR in the preservation era.

2-6-2T Prairie Tanks

Prairie Tanks were introduced by George Churchward, GWR's Chief Locomotive Engineer during this period (see Notable People), in 1903 and saw extensive use across the GWR network of branch lines.

Churchward was keen to maximise efficiency and settled on a number of standard designs for key engine parts so that there could be as much reuse as possible.

The class had several incarnations, from Churchward's original design in 1903 right up until 1949. The later variations after this period were developments of the design by Charles Collett, Churchward's successor.

2-6-2T 4566 in a snow covered Kidderminster Town. ***Small Prairie tanks such as 4566 were used extensively by GWR across their branch lines, including the Severn Valley.***

They break into two main groups, the Large Prairies (numbered 5100, 3150, 5101, 6100, 3100 and 8100) which had driving wheels bigger than 5 feet in diameter and the Small Prairies (numbered 4400, 4500 and 4575) with driving wheels less than 5 feet.

Both types were intended as mixed use engines, equally comfortable pulling suburban passenger services and goods trains. The Small Prairies, such as the SVR's 4566 in the picture below (which was actually built shortly after this period, in 1924 but to the same design), would have been a more familiar sight on the Severn Valley, having been designed primarily for use on branch lines.

2-6-2T ***Large Prairie*** *5164 approaches Bewdley.*

Railmotors

Railmotors were single carriages with a self-contained engine designed mainly for suburban or branch line use.

Steam railcar Enfield in 1849, ***early example of a railmotor.***

The idea of a self-contained carriage wasn't new, as can be seen from the image above they'd been introduced in the mid 19th Century (two years before that image in fact, in 1847) but had fallen out of favour.

The London and South Western Railway reintroduced them in 1902 for one of their branch lines and the GWR borrowed one from them to trial. When it proved successful Churchward designed a Great Western version and from 1903 to 1908 built 99 of them.

As a relatively underused branch line, the Severn Valley was not often at the forefront of new developments but the railmotors were designed with branch line use in mind and as a result the first scheduled service on the line was introduced on 2nd January 1905 only two years after they had been developed. On the same day Foley Park Halt outside Kidderminster was opened to be served by the railmotors.

***GWR Railcar number 93**, built in 1908 it is now based at Didcot.*

Most railmotors were eventually converted to autocoaches which were considered more flexible. An autocoach is a normal engine and coach arrangement but with a coach that has been modified to be able to control the engine. They provided the advantages of a railmotor without the disadvantages.

0-6-0PT Pannier Tanks

Pannier Tanks were another long lived GWR design and a favourite, in their many variations, on branch lines like the Severn Valley. Their small size but powerful engines made them perfect for branch line work and, like the Prairies, they could be found hauling passengers or freight.

The panniers were a development of the late nineteenth century saddle tanks, an example of which, number 2047 Warwickshire, could be seen on display at Kidderminster Railway Museum until it was sent for restoration in 2010.

0-6-0ST 2047 Warwickshire in April 1971. 2047 is a ***saddle tank engine*** *from which the ever popular GWR pannier tanks evolved. Contrast the distinctive curved tank stretching over the boiler with the square, flat topped tank of the Pannier in the photo below. 2047 is now undergoing restoration in Bewdley Down Yard.*

By 1910 no more saddle tanks were being produced by GWR, and those that were already built were being converted to Panniers.

0-6-0PT 7714 taking on water at Kidderminster. 7714 is a ***5700 class Pannier Tank****. Although built in 1930 by Collett she is very similar to earlier pannier designs from the pre-war period. The 5700 class was the single largest group of engines built by the GWR, with 863 being built between 1929 and 1950.*

The tanks on the Pannier offered a lower centre of gravity than the over-the-boiler design of the saddle, allowing the engine to corner faster and increasing safety. However, the main reason for the switch was the introduction of an improved firebox by Churchward, known as the Belpaire, which could not be fitted to the curved tanks of the saddles.

Panniers were a phenomenally successful design with just one class, the 5700, having 863 engines built to it, the biggest number of engines in any GWR class.

Although there were other visitors to the line on occasion, the majority of the engines in regular use, as can be seen from the above, were relatively small when compared to the engines that can be seen in preservation.

Many of today's engines would have seemed out of place on a branch line, at least until the coming of war turned life upside down, which is what we will look at in the next chapter.

CHAPTER THREE: War and Amalgamation

To understand the events leading to the closure of the Severn Valley Railway it is necessary to leave the line for a while and consider the broader situation of the railways across the country.

Apart from one or two tales of the SVR then, this chapter will focus on the wider story of the railways in the first half of the twentieth century.

For those interested only in the Severn Valley I suggest you skip to the end (actually I don't, you'll miss much fascinating information about the role of the railways in war, but you can if you want!)

Although the first and second world wars were obviously very different, the sad truth is that from the point of view of the railways at home they were very similar, so I have provided much of the material in the section on World War I, and for World War II simply highlighted a couple of specific things that set it apart.

THE RAILWAYS IN WORLD WAR I

The Railway Mania described in Chapter One left a legacy of around 130 railway companies covering Britain with a patchwork quilt of track run by different organisations.

Arguably this didn't matter too much in peacetime but in war when it would be necessary to transport troops and material reliably from one end of the country to the other, it would obviously become impractical. As a result the government recognised the need for some sort of coordination as early as the late nineteenth century, forming an Army Railway Council in 1896.

In the early 1900's, however, the body that existed to fill this role was the Railway Executive Committee (REC).

Formed on 5th November 1912, the REC was made up of representatives from the thirteen largest railways with the open goal of liaising with the armed forces to ensure that soldiers and sailors could effectively reach their training grounds and barracks.

Although in 1912 nobody was specifically anticipating the war that would happen two years later, a European war was clearly possible and the government also tasked the REC with preparing.

Just as well they did because when war finally came in 1914, it was triggered quickly and caught many experts unaware. Not so the railways.

So important was rail to the war effort that on the first day of war the government immediately enacted the Requisition of Forces Act 1871 and took control of the railways.

The notice to the public appeared in the same edition of the newspapers that informed them Britain was at war.

"Gazette" £1,000 Free Insurance.

Birmingham Gazette

KAPUTINE

ENGLAND AND GERMANY AT WAR.

GERMAN REPLY TO ULTIMATUM.

NOT SATISFACTORY.

ENGLAND IN STATE OF WAR.

GOVERNMENT TAKE CONTROL OF RAILWAYS.

THE BELGIAN APPEAL.

STATEMENT BY PREMIER.

THE ULTIMATUM.

NEUTRALITY MUST BE OBSERVED.

THE MEN OF THE HOUR.

HOME FLEETS UNDER ADML. JELLICOE.

Sir John French.

THE PERIL OF BELGIUM.

INVASION BY GERMANY.

AEROPLANE BOMBS.

FRENCH FRONTIER TOWN DAMAGED.

At the outbreak of war the **government immediately took control of the railways** *through the* **Railway Executive Committee** *(REC)*

The REC had done their job well and by the time war was declared a detailed mobilisation plan was in place.

By the end of August 1914 the railways across Britain had transported the 120,000 men (and 60,000 horses!) of the British Expeditionary Force and untold equipment to Southampton where they would be transported across the channel to France. Sir Herbert Walker, chairman of the REC said later, "At Southampton, for practically every day of the first three weeks of the war, we handled during a period of fourteen hours no fewer than 73 of these trains… **The trains arrived at intervals averaging twelve minutes**." (Quoted in *British Railways and the War* by Fred McKenzie.)

Sir Herbert Walker*, chairman of the REC.*

As war was declared at the beginning of August it coincided with the holiday season which lead to a conflict between the military's need to get men to their embarkation ports and the public's non-war related plans to go on summer holidays.

The REC instructed the railways to cancel their planned holiday timetables, and ran an advertising campaign to persuade people not to travel unless absolutely necessary.

In September the REC introduced winter timetables early to further reduce the non-military traffic on the lines.

At this point it is important to remember that most people (both the public and in government) expected the war to be over by Christmas, however, that as we know was not the case, and the need to reduce civilian traffic on the lines to ensure that the military could be serviced would continue throughout the war with varying degrees of success. In 1917 the REC went so far as to increase passenger fares by half in an effort to stop people travelling.

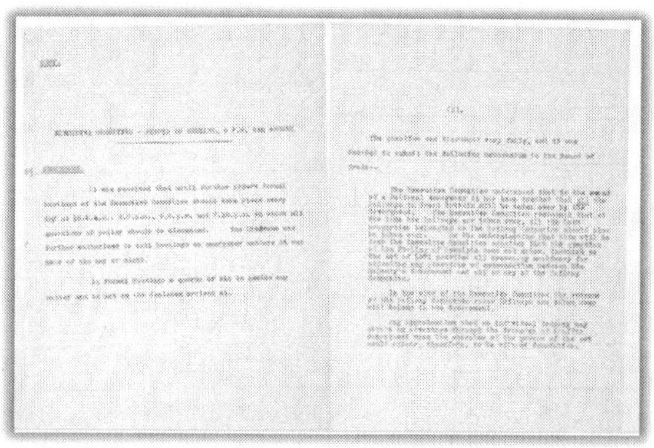

The railways were of paramount importance in the war. In the early days, as can be seen from the official minutes, **the REC met four times a day to ensure the railways were providing everything the mobilisation needed**

Ambulance trains

No discussion of the railways' war work would be complete without mention of the ambulance trains that were key in bringing home the wounded soldiers from the front.

Ambulance trains had been used in earlier wars (including the Boer War as can be seen from the accompanying picture, and indeed the Crimean War where our own Morton Peto had built his military railway) but World War I saw them elevated to new heights.

***Early ambulance train** in the Boer War*

The REC instructed the railway companies to provide updated models. The GWR's design was supervised by Frank Marillier, the Carriage and Wagon Works Manager at Swindon.

A masterpiece of engineering, they included dedicated carriages that served as wards with three tiers of bunks that could hold 144 men lying or transform ingeniously into seating if the patients were able, in which case they could hold even more. They had a pharmacy carriage which contained an operating room, they also had kitchens to prepare meals for the patients and a mess for the officers and nurses.

Patients waiting to board an ambulance train in France.

In the middle of the train the wide double doors used to admit patients on stretchers can be seen. They were one of the most minor alterations made to these amazing carriages.

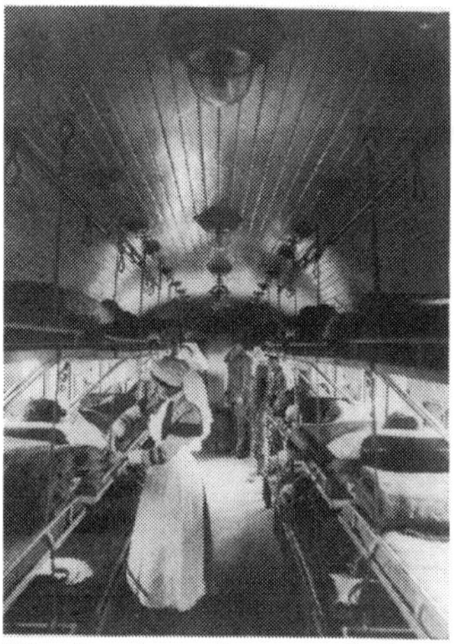

The ***interior of an ambulance train*** *showing the tiered bunks.*

Gone was the beautiful chocolate and cream livery of the GWR, replaced by a khaki colour more appropriate to wartime accompanied by a large red cross on a white background, intended to make it clear that the trains should not be considered a target (something that, as you can imagine, was not completely successful). The trains were operated by members of the Red Cross, the Queen Alexandra's Imperial Military Nursing Service and the Royal Army Medical Corps, and though our story is focused at home, it is worth noting that some of these heroic nurses, doctors and orderlies remained on board for the duration of their service.

By the end of the war the ambulance trains in the UK had transported approximately two million, six hundred and eighty thousand patients to hospital on fifteen thousand special trains, more than six thousand of which ran across GWR lines like the Severn Valley.

Railway Operating Division

The other main area of the war that involved railwaymen in something like their normal occupation was the Railway Operating Division.

Created in 1915 as part of the Royal Engineers its purpose was to run, and in some cases build, the railways that were used on the continent to transport troops to the front. As could be expected, the best people to fill it were recruits from the railway companies.

By the end of the war the number of British railwaymen working for the Railway Operating Division was greater than 24,000.

And of course there were the men who volunteered, or were called up, for service itself. Of the estimated 700,000 men who worked for the railways more than 100,000 enlisted. Railway work was considered an essential home duty so as the war took longer than expected, many volunteers were prevented from going but many served at the front. More than 20,000 of them would lose their lives over the course of the war, including approximately two and a half thousand GWR men. (The approximate number is because the roll of honour below shows 2436, the memorial at Paddington 2524 and the Great War Book 2665, demonstrating the confusion of war).

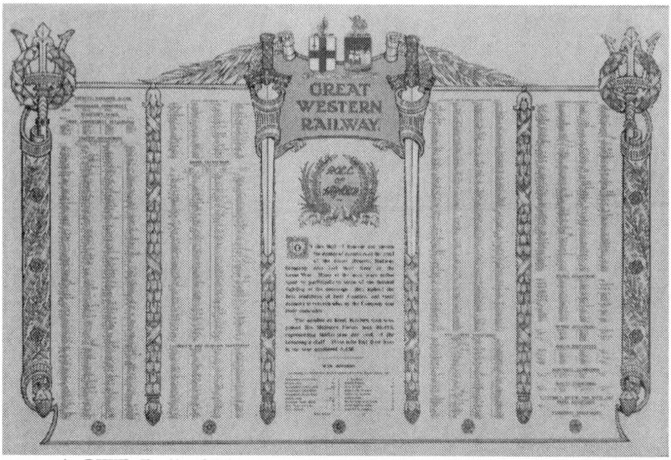

A GWR Roll of Honour *produced by the company and distributed to key stations on its lines. The dedication reads:*

On this roll of honour are shown the names of members of the staff of the Great Western Railway Company who lost their lives in the Great War. Many of the men were called upon to participate in some of the fiercest fighting of the campaign; they upheld the best traditions of their Country, and their memory is revered alike by the Company and their comrades.

The number of Great Western men who joined His Majesty's Forces was 25,479, representing thirty-three per cent. of the Company's staff. Those who lost their lives in the war numbered 2,436.

An unpleasant incident on the Severn Valley

Unfortunately, as the following brief story shows, the perils of war did not remove the usual dangers associated with working on or near a railway.

On 2nd September 1916 at about ten in the evening Private Bernard Brady (number 20420 with the Royal Corps Defence), was walking to his post.

In August 1914, only three weeks after the outbreak of war, and despite being 43 at the time and therefore too old for active service, Brady had signed up for one of the first Pal's Battalions, the 19th Battalion of the King's Liverpool Regiment. By September 1916 he had transferred to the Royal Defence Corps and was guarding the Elan Valley aqueduct.

The Elan Valley aqueduct *as seen from the SVR. The aqueduct brings 360 million litres a day to the people of Birmingham.*

The aqueduct brings 360 million litres of water a day across the 73 miles from Wales to the people of Birmingham and crosses the River Severn midway between Northwood Halt and Trimpley reservoir, passing underneath the railway immediately after crossing the river.

In common with other major waterworks it was recognised as key infrastructure that needed to be protected in case of invasion or sabotage from enemy sympathisers, and it was the Royal Defence Corps' job to do just that.

The 2nd was a Saturday night and Brady was on night duty, making his way along the line to his post to make sure the aqueduct was safe from a German invasion that did not come.

He never reached his post.

A little before ten p.m. an engine driver named Attwood reached Arley on the down line from Bewdley and noticed blood on the front of the engine.

He initiated a search and the horrified searchers discovered Brady's body. He had been hit by the train and decapitated.

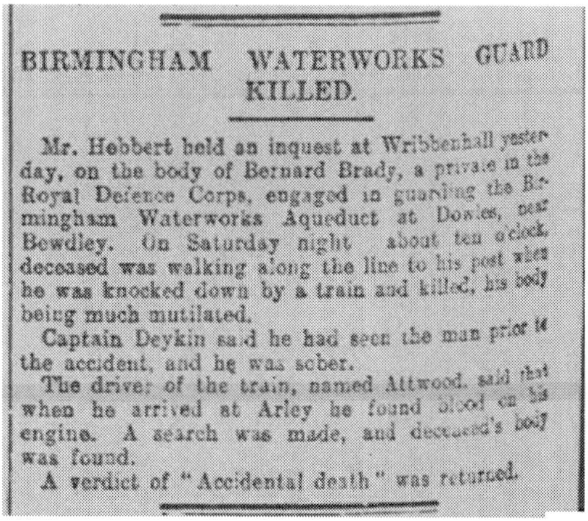

BIRMINGHAM WATERWORKS GUARD KILLED.

Mr. Hebbert held an inquest at Wribbenhall yesterday, on the body of Bernard Brady, a private in the Royal Defence Corps, engaged in guarding the Birmingham Waterworks Aqueduct at Dowles, near Bewdley. On Saturday night about ten o'clock deceased was walking along the line to his post when he was knocked down by a train and killed, his body being much mutilated.

Captain Deykin said he had seen the man prior to the accident, and he was sober.

The driver of the train, named Attwood, said that when he arrived at Arley he found blood on his engine. A search was made, and deceased's body was found.

A verdict of "Accidental death" was returned.

Brady's death as reported in the local press. *The phrase 'much mutilated' here was reported more fully in other papers as 'decapitated'*

His death was ruled an accident.

The records of the inquest are no longer held by the Worcestershire Archives so we cannot know how far Mr A. H. Hebbert the Deputy Coroner questioned Attwood, but we can safely assume it is unusual that a driver would inspect the front of his engine for blood without cause. While obviously an accident, it is hard to believe that Attwood was unaware he had hit something, if not a person. In any case there is no record of Attwood receiving any reprimand on his GWR service record.

Brady was buried in All Saints Church in Wribbenhall, his grave later being recognised by the Commonwealth War Grave Commission.

On 31st July 2019 Brady's three brothers were given a war gratuity of £5,10s, 0d in recognition of his war-time service.

***Private Brady's headstone** in Wribbenhall church.*

BETWEEN THE WARS

After the demobilisation had completed, life on the railways tried to return to normal, however, the REC did not relinquish control until 1921 when the Railways Act 1921 was put in place.

Many people had seen the benefits of the railways acting together through the war and there was a push in several quarters to nationalise them. The Railways Act was the government's response to that push, but it stopped some way short of nationalisation. Even though government was led by the Liberal David Lloyd George, it was a coalition made up of two thirds conservative members who would have opposed outright nationalisation.

Interestingly it was designed by a man, Sir Eric Geddes, who would be the first person associated with the word 'axe' for which Beeching would later become infamous in railway circles.

***Sir Eric Geddes**, yet another fascinating railway man. He can be thought of as the grandfather of British Railways.*
Geddes had a background in railways, having been the general manager of the North Eastern Railway before his distinguished war service. He was the first Minister of Transport and was a keen rationaliser in many areas. The 'Geddes Axe' became famous for cost saving and simplification long before Beeching's (and from our view had the benefit of not being applied to the railways!)

Geddes proposed a reorganisation of the railways that meant they kept their status as independent companies but were amalgamated into a much smaller number of large companies. In that way he believed he would do away with the pre-war state of competition between more than a hundred smaller companies and retain the benefits that had been seen by them working together through the war.

The Act, known informally as the Grouping Act, brought together the vast majority of the railways in the country at that time into four main companies:

1. *the Southern Railway – SR;*
2. *Great Western Railway – GWR;*
3. *London, Midland and Scottish Railway – LMS; and*
4. *London and North Eastern Railway – LNER.*

As can be seen, the GWR was the only company to survive the Grouping with its identity intact. The Severn Valley, as a branch line of the GWR, is not mentioned explicitly in the Act.

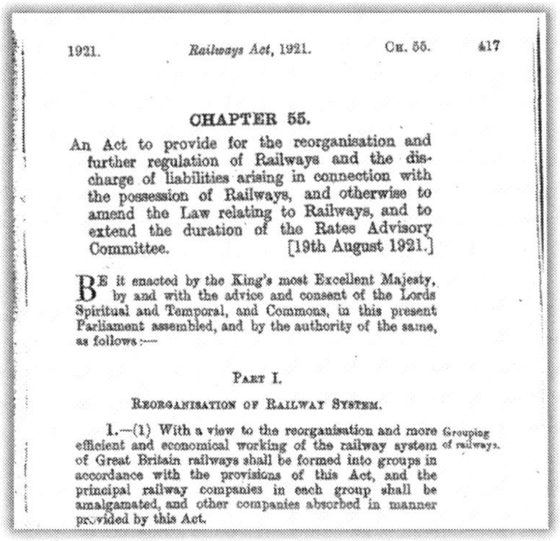

1921. *Railways Act*, 1921. Ch. 55. 417

CHAPTER 55.

An Act to provide for the reorganisation and further regulation of Railways and the discharge of liabilities arising in connection with the possession of Railways, and otherwise to amend the Law relating to Railways, and to extend the duration of the Rates Advisory Committee. [19th August 1921.]

BE it enacted by the King's most Excellent Majesty, by and with the advice and consent of the Lords Spiritual and Temporal, and Commons, in this present Parliament assembled, and by the authority of the same, as follows:—

PART I.

REORGANISATION OF RAILWAY SYSTEM.

1.—(1) With a view to the reorganisation and more efficient and economical working of the railway system of Great Britain railways shall be formed into groups in accordance with the provisions of this Act, and the principal railway companies in each group shall be amalgamated, and other companies absorbed in manner provided by this Act. Grouping of railways.

***A copy of the Railways Act 1921**, the Grouping Act that formed the main parts of what would later become British Railways.*

Back on the Severn Valley

The Grouping of the railways had relatively little impact on the Severn Valley Railway. As a branch line well in the heart of GWR-land it was largely unaffected. In fact, the biggest change on the line during this period was probably the opening of a number of new halts.

Never particularly excessive, demand had started to fall further in 1920 and GWR saw halts as the solution. They were cheap to run and enabled more fare paying passengers to join the line.

Travelling down the line (i.e. towards Shrewsbury and away from London) the halts were:

Burlish – between Stourport and Bewdley. It was opened on 31st March 1930 with one platform and a pagoda style shelter (like the one that can be seen at Northwood today).

Northwood – between Bewdley and Arley, opened on 17th June 1935. Northwood is described in more detail in Chapter Eight, p. 157.

Jackfield – between Coalport and Ironbridge, opened on 3rd December 1934.

Cound – between Cressage and Berrington, opened on 4th August 1934.

And on the Kidderminster Loop, *Foley Park*, which was originally opened in 1905 as we saw in Chapter 2, was moved from the south side of the line where it had been put to serve the local community, to the north side in 1925 when the West Midlands Sugar Company opened a sugar beet refinery at Foley Park. (West Midlands Sugar became part of the British Sugar Corporation in 1936 following yet another government amalgamation).

Whilst discussing halts it is worth noting that the other two (or three depending how we count) on the line both fall outside of this period.

Rifle Range was opened on the Kidderminster Loop on 12th April 1905 and, contrary to the general approach during this period, was closed in October 1920.

Eardington was originally opened as a station but later downgraded to a halt by British Railways in 1952.

Alveley was opened later to serve the Highley Mining Company coal miners at Alveley colliery. Sources differ as to when it opened, some dating it from the opening of the sidings themselves which definitely took place on 30th January 1939, others not until 1944.

Eardington Halt in 2011. *The halt hasn't been used in preservation since 1982 but has been maintained by a group of preservationists. At the time of writing (2019) a plan is underway to reopen the halt as a station in connection with the Astbury Hall stately home and golf course (former home of Judas Priest lead guitarist, Ken Downing).*

Sadly, the experiment to increase local custom should probably be considered a failure. The line was never going to be busy by national standards and all of the halts were subsequently closed and, with the exception of Northwood, do not form part of the reopened line.

THE RAILWAYS IN WORLD WAR II

In World War II the road network was still in its infancy (the first motorway was not built until 1958) so the railways once again took on their role of coordinating the mobilisation.

The Railway Executive Committee was reformed on 24th September 1938 under the chairmanship of Sir Ralph Wedgwood, great-great-grandson of Josiah Wedgwood, the founder of the famous pottery.

Sir Ralph Wedgwood Bt, *chairman of the Railway Executive Committee 1938 - 1942.*

Sir Ralph was chairman of LNER and had the distinction of having two 4-6-2 A4 Pacifics named after him. The first, 4469, was destroyed by German bombs in April 1942 and a second, number 4466, was renamed for him.

The REC took control of the railways on 1st September 1939, days before the declaration of war on Germany.

In many respects the role of the railways in World War II was the same as those we have described in World War I, so we will only look here at some of the key elements that stand out.

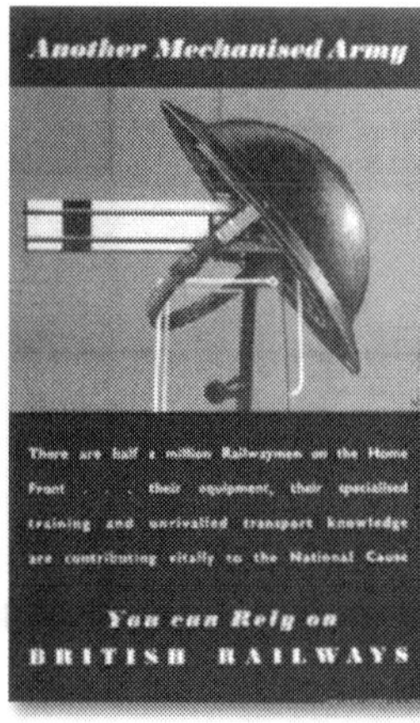

Posters coordinated by the REC during World War II. *At this stage 'British Railways' did not mean the company that was formed later but 'the railways in Britain', i.e. the Big Four*

Evacuation

More advanced aircraft meant that the threat of aerial bombardment was significantly greater in the second world war than it had been in the first. As a result the decision was made to evacuate as many people as possible from affected areas (primarily London and the south). Given the plan's understandable attention on children, the evacuation was known as Operation Pied Piper.

On 24th August teachers were called back from their holidays early and schools opened the next day.

In the first three days of the evacuation more than one and a half million people were evacuated. Most passed through the railways.

Children being evacuated from Waterloo *passing soldiers on their way to embarkation. The luggage label that can just be seen around the neck of the little girl at far right would have listed her name, address, school and her destination.*

The railways continued to evacuate people throughout the war but shifted to a 'Trickle' plan with trains leaving the cities daily or weekly depending on the phase of the war.

Ministry of Health poster *displayed in the London Underground to encourage evacuation.*

REC's Top Secret Underground Headquarters

In the run up to war the REC headquarters were at Fielden House on Great College Street but the same appreciation of the likelihood of bombs being dropped that led to the evacuations meant that it was obvious that a more bomb proof location would be required.

In April 1939 approval was given for them to convert a disused tube station, Down Street in Mayfair, to their new headquarters.

The street entrance to the ***Down Street top secret headquarters of the REC.*** *Two uniformed policemen would guard this door 24 hours a day.*

Being a tube station there were no offices already in place, the only available space was the platform itself. There was also the small problem of the Piccadilly line trains that ran through the disused station on a regular basis. But that didn't stop them.

One night after the tube had stopped running, workmen from the London Passenger Transport Board descended on the station and walled the platform off, hiding it from sight of anyone travelling on the line.

The conversion could then continue in complete secrecy. As well as a series of meeting rooms and offices, the final building would include a bomb proof lift; a state of the art telephone exchange and radio equipment; a back-up generator that automatically kicked in if the main power was lost; bedrooms for staff to sleep on site; officers mess (complete with wood panelling and wallpaper!); and most importantly a control room covered in maps of the railway network across the country to allow the staff to coordinate everything (though in the digital age it should be remembered that these were paper maps, real-time update of the 'display' meant receiving a phone call and someone adding or removing pushpin flags as necessary!)

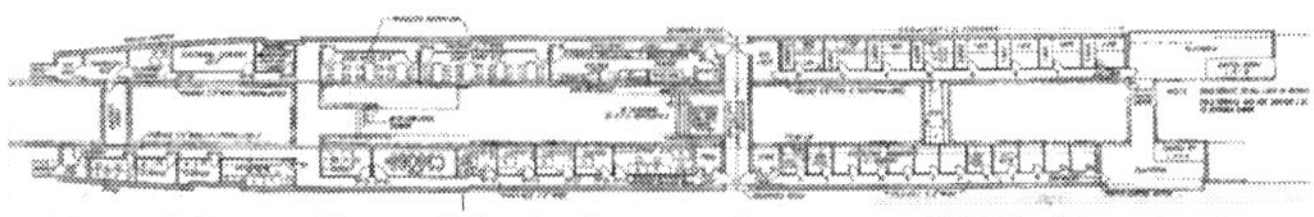

Plan of Down Street *following its conversion to secret REC offices, showing the shortened platforms where REC members would stop passing trains to join the driver unknown to the passengers*

Two uniformed police officers stood guard on the door to the street, but most access was via the underground itself. The LPTB had built short, secret platforms with traffic control buttons which the REC staff could use to stop a passing train and join the driver in the cab without the passengers being aware before traveling to Green Park or Hyde Park Corner stations and emerging into the crowd.

After the REC were installed Sir Ralph's brother Josiah, a member of the cabinet, suggested to Churchill that it might be a useful (and safe) place to conduct War Cabinet meetings until the construction of the bunker at the Cabinet War Rooms was completed.

According to the diaries of his secretary Churchill was equally delighted with the secure facilities and the catering at Down Street! The food was provided by the railway hotels and included caviar, champagne and brandy. Churchill's secretary's diary for 30th October 1940 says that Churchill and Ernest Bevin ate (and evidently drank) at Down Street and when leaving were drunk enough to struggle to get the lift to work. When they finally got outside Churchill only narrowly avoided getting arrested after having an argument with a policeman about how bright his car sidelights were.

The REC in their underground committee room in Down Street. *Note the curved walls of the converted tube platform.*

BACK ON THE SEVERN VALLEY, PT. 2

World War II saw a number of changes to the local area that had an impact on the Severn Valley Railway.

Military Presence

Over the course of the war three military bases opened along the line.

RAF Bridgnorth – was used as a training ground for RAF recruits. It was in operation throughout the war and beyond with tens of thousands of recruits doing their initial training there. Many of them were brought in by train from RAF Cardington, the lucky ones detrained at Bridgnorth station and were brought to the camp in open backed truck, others being deposited at a halt and being made to run across the fields to reach their new home. Based on the location this was probably Eardington, and would technically have been an unmanned station at that point. From August 1941 until September 1942 it trained recruits for the Women's Auxiliary Air Force. The site is now Stanmore Country Park.

Spitfires c. 1941. *This picture was taken in RAF Castletown but shows the iconic type of British WWII aircraft that recruits would have trained for in RAF Bridgnorth, and stored parts for in RAF Hartlebury.*

RAF Hartlebury – opened in 1938 was the base of No. 25 Maintenance Unit. Although most of the personnel on site were civilians, RAF Hartlebury was used for just about everything the RAF could need to fight the war from medical supplies to spare parts for Spitfires and fighter engines.

Burlish Camp – In 1944 two hospital camps, the 114th and 297th General Hospitals, were opened by the US Army at Burlish just outside Bewdley. Between them they had 1254 staff and treated more than 12000 patients before closing in June 1945.

Most of the injured arrived at Stourport station but some came via Kidderminster.

On 31st May 1944 General Patton visited Burlish Camp where he gave his most famous speech.

General Patton, commander of US 3rd Army.
Patton gave a version of his famous speech at Burlish Camp overlooking the Severn Valley Railway.
"You are not all going to die," he said slowly. "Only two percent of you right here today would die in a major battle. Death must not be feared. Death, in time, comes to all men. Yes, every man is scared in his first battle. If he says he's not, he's a liar… The real hero is the man who fights even though he is scared…"

'Upgrade'

That military presence inevitably led to a huge increase in traffic along the line and in light of that, and its use as a diversionary route for the major Midlands cities, the classification of the line was upgraded.

Until then GWR had classed the Severn Valley as a light route which meant it was restricted to lower weight engines. The new need resulted in it being reclassified as a 'blue' route, allowing larger engines up to a maximum of 17 tons 12 cwt axle loading to use it (although some had speed restrictions applied at various points such as Victoria Bridge).

In practice of course, there were occasions when both the speed and weight restrictions were not applied either by mistake or by being overlooked due to wartime need.

Another sort of line

The final point of note before we leave the war period altogether is the part the Severn Valley was to play in the event of German invasion.

In June 1940 in response to an order from Field Marshall Ironside, Commander in Chief of the Home Forces, plans were drawn up to create a series of 'stop lines' across the country.

A memo from Southern Command at the time explains the point of a stop line:

> *"The immediate object is to divide England into several small fields surrounded by a hedge of anti-tank obstacles which is also strong defensively, using natural accidents of the ground where possible. Should A.F.V.s or airborne attacks break into the enclosures the policy will be to close the gate by blocking the crossing over the obstacle and let in the 'dogs' in the shape of armoured formations, or other troops, to round the cattle."*

The Western Command Stop Line No. 1 ran along the River Severn and the Severn Valley Railway over Victoria Bridge formed a key part of it.

The local Home Guard, who were stationed in the sailing club at Trimpley, dug trenches at the base of it and installed railblocks on the south end. The railblocks were simple concrete blocks with holes in the top placed on either side of a railway line. In the event of invasion the Home Guard would have placed strong steel bars through the concrete blocks to cut off the railway and stop any advancing army from using the railway.

On the Kidderminster loop the tunnel entering Kidderminster was also highlighted as a vulnerable point and protected by barbed wire entanglements and placed under armed guard.

As Malcolm Atkin, former head of the Historic Environment and Archaeology Service for Worcestershire, has pointed out, the sad truth was that the anti-tank islands formed in this way were positioned around attractive targets and were actually chosen not just to halt the German advance but to draw them in and hold them in place until the regular army had time to regroup and defend the homeland. The role of the Home Guard was to have harried the enemy but they and the inhabitants of the islands were acceptable casualties of this plan. Both Kidderminster and Worcester were designated islands.

The barbed wire is obviously long gone, but the railblocks can still be seen to this day by the side of the track near Victoria Bridge, a reminder of the very real possibility of invasion.

CHAPTER FOUR: Nationalisation, Modernisation and Closure

All of the work carried out in support of the two wars was amazing but it took a toll on the railways themselves. During the wars the maintenance of the lines, engines and stock fell away to the barest minimum, and this stored up problems for the period immediately afterwards. This underinvestment and a lack of an appropriate response to the growing use of private cars, ultimately led to the closure of many lines, including the Severn Valley Railway.

Dr. Beeching rightly gets the popular blame for the closure of many of the railways in the 1960s but where the Severn Valley is concerned the story isn't quite that straightforward.

This chapter will describe three shots which were fired towards the railway and that might have ended in its closure:

- The Modernisation Plan
- BR's internal rationalisation
- The 'Beeching Axe'

What is clear is that between the Modernisation Plan's removal of steam, British Railway's own drive for rationalisation of the network and then the brutal axe of the Beeching Report, there was no way a line as little used Severn Valley was going to survive. However, before any of these could have an impact on the line, it had to contend with nationalisation.

NATIONALISATION

When Clement Atlee came to power on 26th July 1945 his Labour government were clear about their desire to nationalise many industries, and the railways were high on the list.

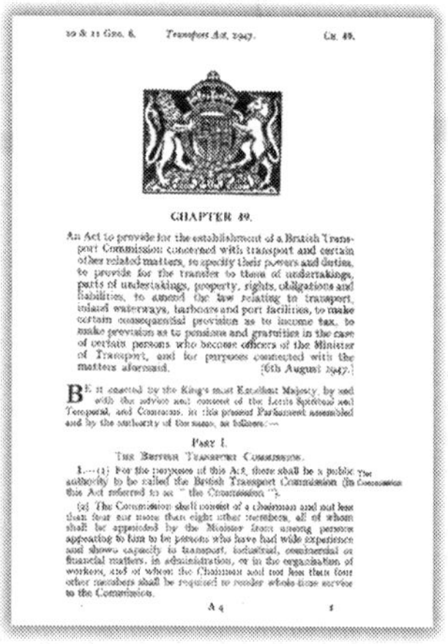

10 & 11 Geo. 6. *Transport Act, 1947.* Ch. 49.

CHAPTER 49.

An Act to provide for the establishment of a British Transport Commission concerned with transport and certain other related matters, to specify their powers and duties, to provide for the transfer to them of undertakings, parts of undertakings, property, rights, obligations and liabilities, to amend the law relating to transport, inland waterways, harbours and port facilities, to make certain consequential provision as to income tax, to make provision as to pensions and gratuities in the case of certain persons who become officers of the Minister of Transport, and for purposes connected with the matters aforesaid. [6th August 1947.]

BE it enacted by the King's most Excellent Majesty, by and with the advice and consent of the Lords Spiritual and Temporal, and Commons, in this present Parliament assembled and by the authority of the same, as follows:—

PART I.

THE BRITISH TRANSPORT COMMISSION.

1.—(1) For the purposes of this Act, there shall be a public authority to be called the British Transport Commission (in this Act referred to as "the Commission"). The Commission

(2) The Commission shall consist of a chairman and not less than four nor more than eight other members, all of whom shall be appointed by the Minister from among persons appearing to him to be persons who have had wide experience and shown capacity in transport, industrial, commercial or financial matters, in administration, or in the organisation of workers, and of whom the Chairman and not less than four other members shall be required to render whole-time service to the Commission.

A 4 1

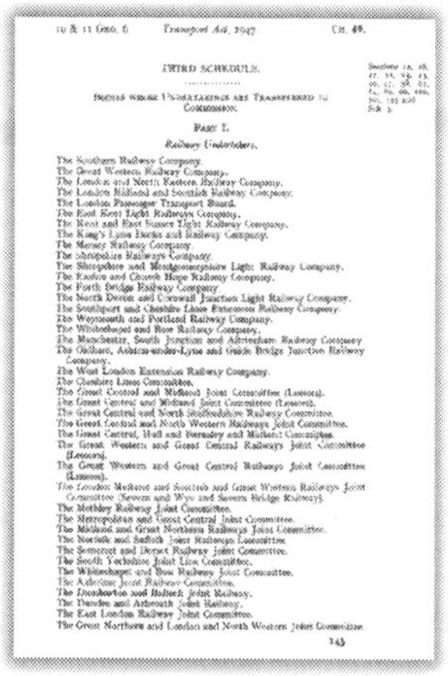

10 & 11 Geo. 6. *Transport Act, 1947.* Ch. 49.

THIRD SCHEDULE.

BODIES WHOSE UNDERTAKINGS ARE TRANSFERRED TO COMMISSION.

PART I.

Railway Undertakers.

The Southern Railway Company.
The Great Western Railway Company.
The London and North Eastern Railway Company.
The London Midland and Scottish Railway Company.
The London Passenger Transport Board.
The East Kent Light Railways Company.
The Kent and East Sussex Light Railway Company.
The King's Lynn Docks and Railway Company.
The Mersey Railway Company.
The Shropshire Railways Company.
The Shropshire and Montgomeryshire Light Railway Company.
The Forth Bridge Railway Company.
The North Devon and Cornwall Junction Light Railway Company.
The Weymouth and Portland Railway Company.
The Whitechapel and Bow Railway Company.
The Manchester, South Junction and Altrincham Railway Company.
The West London Extension Railway Company.
The Cheshire Lines Committee.
The Great Central and North Staffordshire Railway Committee.
The Great Central and North Western Railways Joint Committee.
The Great Western and Great Central Railways Joint Committee (London).
The Methley Railway Joint Committee.
The Metropolitan and Great Central Joint Committee.
The Midland and Great Northern Railways Joint Committee.
The Norfolk and Suffolk Joint Railways Committee.
The Somerset and Dorset Railway Joint Committee.
The South Yorkshire Joint Line Committee.
The Whitechapel and Bow Railway Joint Committee.
The Dundee and Arbroath Joint Railway.
The East London Railway Joint Committee.
The Great Northern and London and North Western Joint Committee.

143

The Transport Act 1947 *subsumed the Great Western and with it the Severn Valley Railway. GWR is the second company on the list at right.*

The Transport Act 1947 created the British Transport Commission (BTC) and with it a new Railway Executive (distinct from the earlier REC from the war period).

British Railways was born on 1st January 1948 as the public facing name of the new Executive, and with that the Great Western Railway, and more than a hundred years of history ceased to exist.

On the same date the Severn Valley Railway, as part of the GWR, became part of British Railways' Western Region.

***British Railways was formed on 1st January 1948** and the SVR became part of the Western Region.*

The infamous "flying sausage" logo was used in stations but never on locos.

On the Severn Valley, little immediate impact was felt, however, it did bring it within the realm of decision makers who were to ultimately bring about its closure.

The 'lion on a unicycle' logo applied to British Railways' engines.

It was designed by Abram Games based on a sculpture by Cecil Thomas for the British Transport Commission.

The lion on the design was reversible and was always painted facing the front on steam engines.

THE MODERNISATION PLAN

Seven years after British Railways came into being it was apparent that the under investment and lack of maintenance during the first half of the twentieth century, when the railway companies were understandably distracted by the war effort, meant that significant changes would be needed.

In December 1954 the BTC published a report, *Modernisation and Re-Equipment of the British Railways*, known (despite being released in 1954) as the 1955 Modernisation Plan.

Over 36 pages it set out a plan to reinvigorate the railways, calling for an investment of £1.2 billion, or almost £31bn in

2019 (although half of that amount would be required simply for normal maintenance).

As far as steam fans are concerned much of the blame for the decline should lie here not with Beeching.

The report spelled out that steam did not have a future on British Railways.

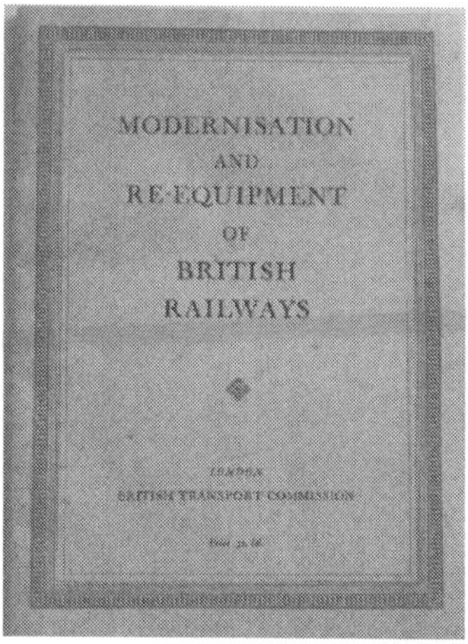

The 1955 Modernisation Plan.

This small, 36 page booklet called for the elimination of steam and the introduction of DMUs across branch lines like the Severn Valley Railway.

The Plan states as its first goal the upgrade of the main track and introduction of improved signalling, however, the second goal was that:

> *" ... **steam must be replaced as a form of motive power,** electric or diesel traction being rapidly introduced as may be most suitable in the light of the development of the Plan over the years; this will involve the electrification of large mileages of route, and the introduction of several thousand electric or diesel locomotives..."*

The report sought, and was granted, £345 million (c. £8.5 billion in 2019) to be spent on the replacement of steam. In a section focused on the electrification of the main line it said,

> *"By these schemes there will be a* ***virtual elimination of steam*** *from important areas of the country."*

Finally, when discussing the expansion of diesel multiple-units to operate branch lines, the Western Region (of which the SVR was of course part) was specifically called out,

> *"... on the Western Region it is intended to eliminate all steam working beyond Newton Abbot..."*

Excluding the Vale of Rheidol narrow gauge line, the replacement of steam on the old GWR lines was complete by 1965.

It is widely agreed that the Modernisation Plan was a failure. The move away from steam was a complete turn around from British Rail's previous position, and as a result while appropriate diesel options were no doubt available many of those that were chosen did not perform particularly well. In fact many of them were withdrawn from service before the steam engines they were supposed to be replacing.

CLOSURE

The failure of the Modernisation Plan to bring about the promised return of traffic from the roads to the railways (with a resulting increase in revenue), meant that the

Transport Commission's attention switched to reducing expenditure. And that meant cuts.

Dr Richard Beeching, *chairman of the BTC 1961-63 and BR 1963-65…*

…and his report that slashed branch lines across Britain.

In March 1961 Dr Richard Beeching, was made chairman of the Transport Commission and at the same time it was announced that he would also be the first chairman of the new British Railways Board when it replaced the Transport Commission in January 1963.

Despite having no experience of running a railway he was an experienced businessman with a distinguished record in both war and peacetime, and he was brought in to bring the railways back into profitability.

He immediately embarked on preparation of a two year long study of the railways with a view to identifying targets for the most drastic set of cuts in railway history, which would become known as the 'Beeching Axe'.

With so much attention focused on the Beeching Axe over the years (including even the nineties comedy series, *Oh, Dr Beeching!*), it would be natural to assume that it was to blame for the Severn Valley Railway's closure. As we shall see shortly that view is not quite accurate but, just in case it looks like I am underestimating the impact of the Axe (or, heaven forbid, defending Beeching!) it is worth highlighting the sweeping nature of the cuts he brought in.

The Beeching Report

On 27th March 1963 the British Railways Board, of which Dr Beeching was now chair, published a 162 page report (including the maps) called *The Reshaping of British Railways.*

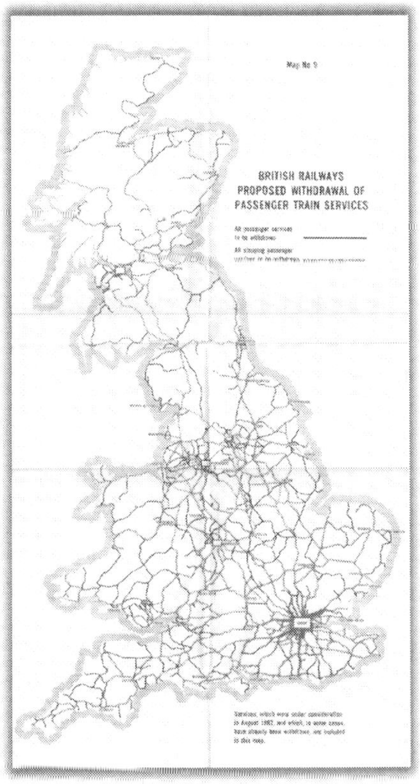

Map showing ***lines to be closed*** *under Beeching*

Much as we might dislike it, and even disagree with the conclusions, the need for a systematic study and the attention to detail is undeniable. The Modernisation Plan had been produced based on a belief that moving to diesel and electric traction would be enough to make the railways profitable and no real thought had been given to the layout or usage of the lines.

Beeching sent analysts out across the network to record the volume of freight, number of passengers, incoming receipts and cost of running each line and station.

Sadly from our perspective the Severn Valley Railway did not fare well by any of these measures. In the final report the results were grouped into categories and the Severn Valley was ranked in the lowest category in each measure.

The report was damning about the state of the railways in general, but Beeching saved his greatest criticism for branch lines or 'stopping services'. It is worth quoting a couple of passages here to get a feel for what the Severn Valley and lines like it were up against.

> *"Thus, in the passenger field,* ***stopping-trains are by far the worst loss maker.****"*

Beeching provided statistics that showed that in 1961 if they had been considered on their own, branch lines lost almost £56m (£1.1bn in 2019 terms).

Nor were they well used:

> *"... are known to be very lightly loaded and to run, very largely, on routes* ***which carry little traffic*** *of any kind."*

And,

> *"Amongst traffics,* ***stopping passenger services are exceptionally poor...****"*

And in a form of circular logic which to his credit doesn't appear elsewhere in the report, he also objected to them on the grounds that they were:

"one of the main causes for the continued existence of many of the small and uneconomic stations."

Which is to say, if we didn't have these branch lines we wouldn't need the pesky stations on them either!

The final nail in the coffin though was the money, Beeching pointed out that closing the branch line services he had identified would save £18m a year (£357m today).

To save that sum he listed 2363 stations and halts which would be closed. The list covers 26 pages of the report and with the benefit of hindsight makes for very sad reading. But Beeching did not intend to hang about to make it happen:

"As soon as the required procedure permits, it is desired to withdraw those passenger train services which are clearly uneconomic."

Beeching believed that the branch lines would be replaced by bus services and private car ownership and envisioned the railway instead acting as the main transport arteries of the country.

We can see now that he was as mistaken as the Modernisation Plan before him. He missed the vital community role played by the railways in rural areas and failed to foresee that the role he was intending for the trunk lines would be filled by the developing motorway network.

However, as indicated earlier, while there is no doubt that Beeching Report would have called for the Severn Valley to be closed, the truth is it had already been earmarked for closure before the report was even published.

Rationalisation

Before the Beeching Report was published British Rail had already begun a programme of rationalising lines.

In October 1961 inspectors visited the Severn Valley to record passenger numbers and in June 1962, almost a year before the Beeching Report was published, British Railways announced that passenger services would be withdrawn between Shrewsbury and Bewdley.

0-6-0 3204 at Shrewsbury Station in June 1962. British Railways **announced the same month this picture was taken that the Severn Valley line from Shrewsbury to Bewdley would be closed.** *(3204 was withdrawn less than a year later in March 1963 but her sister engine 3205 would pull the first passenger train on the reopened SVR on 23rd May 1970)*

So while we can blame Beeching for the closure of the Severn Valley Railway, he was after all the BR chairman at the time, it is not quite as straightforward a story as for many branch lines.

Nonetheless, the Severn Valley, under the accurate but less romantic name of "Shrewsbury – Hartlebury", did appear in the Beeching report. It can be found on page 130 under section 6, "*Passenger Services under Consideration for Withdrawal before the Formulation of the Report*".

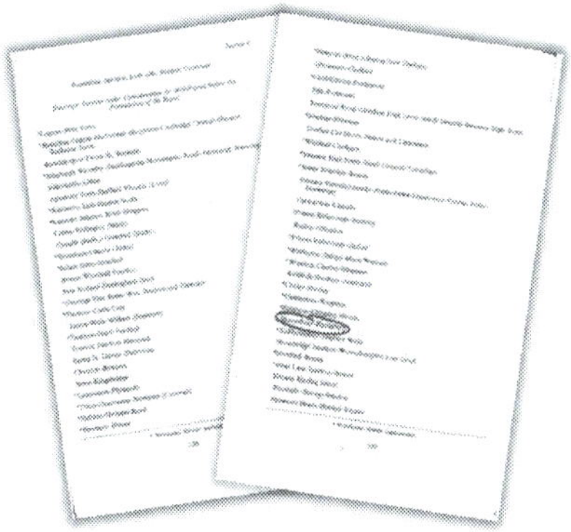

The entry confirming closure of the Severn Valley line in the Beeching Report.
Despite the weaselly words 'under consideration' in the title, the closure had already been announced and the last train would run six months after the report was published.

An appeal was lodged on the basis that closing the line would cause hardship to the people it served, but after many months it was rejected and it was confirmed that the line would close on 9th September 1963.

Between the announcement and the closure there were two incidents that must have felt like they were adding insult to injury for the staff working the line, but in reality are just examples of the bureaucracy of large organisations.

First, in September 1962 Beeching's analysts worked the line to collect information for their report. As we have seen the Severn Valley was not reflected well in that analysis but we can only imagine how remarkably pointless it felt for the staff involved.

Then, on 1st January 1963 the boundaries between British Railways regions were withdrawn and the Severn Valley Railway transferred from the Western Region to the London Midland, meaning that the ultimate notice of closure came from them.

27th August 1963, less than a fortnight before the line closed, what appears to be a Class 119 DMU leaves Coalport station for Hartlebury…

… and then on the evening of 8th September 1963, **the last train leaves Bridgnorth** *pulled by 0-6-0PT 9624 and 4665. Also usually a diesel service, the panniers had been put on specially.*

A reduced passenger service remained in place between Hartlebury, Bewdley and Kidderminster until British Rail finally closed that too in January 1970.

By then though a dedicated group of steam enthusiasts had already decided to take matters into their own hands and the next chapter of the line had begun.

CHAPTER FIVE: Saving the Line

A sense of excitement

In 1965, two years after the line between Shrewsbury and Bewdley was closed to passengers, a meeting took place that would end up saving the line and, in time, lead to the opening of one of the region's most popular tourist attractions.

Keith Beddoes, a Kidderminster engineer with Bradley and Turton Ltd, had heard about the opening of the Bluebell line, the first preserved standard gauge railway to run passenger services and wondered if something similar could be achieved for his local line.

0-60T A1X No. 55 Stepney on the Bluebell Line.

The Bluebell was the first standard gauge preserved line to carry passengers and helped give Keith Beddoes the idea to accomplish the same thing on the SVR.

He discussed it with colleagues and a group of them met at Tony Tuite's house to consider the idea. By the end of the meeting they had convinced themselves that it was feasible enough to see if anybody else was interested.

They contacted a local paper, the Kidderminster Shuttle, and invited anybody interested to join them for a meeting at the Coopers Arms pub, on 6th July 1965.

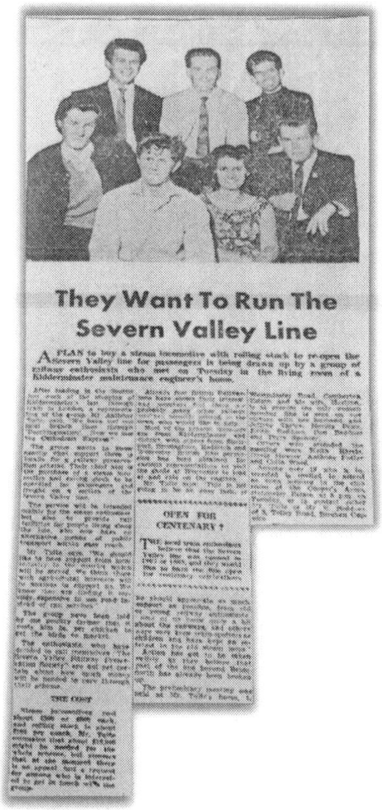

They Want To Run The Severn Valley Line

A PLAN to buy a steam locomotive with rolling stock to reopen the Severn Valley line for passengers is being drawn up by a group of railway enthusiasts who met on Tuesday in the living room of a Kidderminster maintenance engineer's home.

Article in the Kidderminster Shuttle on 2nd July 1965 *calling for anyone interested in saving the Severn Valley Railway to meet at the Coopers Arms.*

Whenever they discussed it with the press they said that approximately fifty people attended, but according to Christopher George in Magner's 'Severn Valley Railway Memories', it was more like thirty five.

About thirty five people attended the meeting and despite most of them being strangers the feeling of possibility was obviously catching. Keith Beddoes said later in 2015 on the fiftieth anniversary:

"We were just a group of ordinary people who didn't know each other, yet there was such a sense of excitement between us at what we had taken on."

That night they formed the Severn Valley Railway Society and agreed to carry out an inspection of the line.

The next weekend, Sunday 11th July, some of them met at Bridgnorth to walk the line to Hampton Loade. They already knew that a year earlier the track had been taken up from Buildwas to a point just north of Bridgnorth, so when they saw the signals and installations had been removed they were not overly surprised.

Walking the line in 1965. *Note the weeds growing along the track!*

What came as more of a surprise was when George Thorpe, who still ran the Refreshment Room as a bar for a small number of locals, mistook them for intruders and pointed the business end of a shotgun at them while demanding to know what they were doing!

After everybody had calmed down, they explained they were thinking of buying the railway and a sceptical George let them into the station buildings to look around.

Another surprise followed when they discovered that he had been using the one of the rooms as a pen for his pig! Nevertheless they left happy, having got the lay of the land.

When they revisited a fortnight later on 25th July, however, they were horrified to discover that since their previous visit British Rail (as British Railways had started to call itself by this time) had also started dismantling the line and buildings at Bridgnorth.

That would have killed their idea before it even got off the ground, but being a Sunday there was nothing that could be done about it so a sleepless night was had by all. The next day John Garth (later Chairman of the Society, and later still Chairman of each of the main organisations of the SVR: the SVRA, the Holding Company, the Guarantee Company) sent a telegram to British Rail asking them to stop any more demolition until they could discuss buying the line.

Nothing was heard from British Rail for two nail-biting days.

Finally at 5pm on Wednesday the volunteers received a reply saying that British Rail were prepared to discuss it and that the dismantling would be paused.

Sadly it was too late to save the signal box but the intervention came in time to stop the dismantling of the line and the footbridge between the platforms.

Working party clearing Bridgnorth station. *The bridge, which barely escaped being demolished and still stands today, can be seen in the background.*

A brief story on the local BBC news website shows how close the whole thing came to never happening.

After the line was closed in 1963 a number of local children used to play on the track at Bridgnorth (a fact that, even without trains running on it, would no doubt cause any Health and Safety conscious modern parent to have kittens!)

One day the children were playing on the line between the station and Bridgnorth tunnel (visible in the picture on the next page) and saw a workman using a machine they had never seen before. He would roll it along the line and then move a little engine from each side of the machine to the other, then roll another couple of feet, stop and run the machine on either side again.

They plucked up the courage to ask him what he was doing, and he answered that he was using the machine to undo the bolts that held the rails together so that he could take up the track.

Satisfied, the kids went back to their game and left him to it while he made his way along the track, removing the bolts as he went.

After a short while he went to take a break, and when he came back the children were confused to see that he seemed to be working the machine back along the parts he had already done. They went and spoke to him again and he explained that he was refastening the bolts he had spent hours undoing because somebody had just bought the station!

Bridgnorth tunnel. *The SVR do not own the tunnel and it can no longer be reached from Bridgnorth station. (Keith Beddoes closest to camera.)*

The first section of line to be saved ran from Bridgnorth station to the sidings at Alveley colliery, the site of the present day Country Park Halt (see Chapter Eight, p. 142).

They could not go further at this point, of course, because British Rail were still running trains to Alveley (although passenger trains had been withdrawn beyond Bewdley, freight trains continued to service the colliery until it closed in 1969).

The original plan was to rent the line but British Rail policy would not allow that, so instead the idea of buying it outright was raised. British Rail were, however, prepared to allow the Society to rent the station while they tried to raise the necessary money to buy it.

K4, 3442 2-6-0, The Great Marquess, in March 1967 at Tyseley shed. 18 months earlier it pulled an SLS special that ran to the edge of the SVR.

On 19th September the Stephenson Locomotive Society ran a "Restored Engines" railtour that came tantalisingly close to the preserved line.

The LNER K4 2-6-0, 3442, *The Great Marquess*, owned by Viscount Garnock and soon after to be a resident on the Severn Valley, pulled a railtour from Birmingham Snow Hill to Worcester Shrub Hill. It was picked up there by tank engines 0-6-0T 1420 and 2-6-2T 4555 and run along the original Severn Valley line via Hartlebury and Bewdley as far as Alveley Sidings.

2-6-2T 4555, seen here at Witham, Somerset in May 1958. In 1965 no. 4555, along with 0-6-0T 1420, pulled the SLS special to Alveley Sidings at the very edge of what was to become the preserved SVR.

As well as giving the volunteers the thrill of a taste of what they were working for, the railtour also gave them an opportunity to recruit more supporters and they handed out leaflets to the three hundred or so passengers on the tour.

The very next week, on 26th September, the first working party started clearing the line.

Negotiating with British Rail

At first the Society had aimed to raise £45,000 a colossal sum equivalent to approximately £775,000 in 2019, with ten percent of that to be paid to British Rail by 4th November if the line was going to be spared.

However in October 1965, after a valuation was carried out on the buildings and line, the Society made an offer of £25,000 (or about £430,000 today). At the same time they asked for the deadline for the deposit to be extended to give them time to find the money.

Another nail-biting wait ensued while British Rail considered the offer.

In fairness to British Rail, we must remember that although the SVR Society wasn't the first group to make a request like this, it was a hardly an everyday occurrence. This was the very earliest days of steam preservation and the idea of a bunch of amateurs trying to buy a railway line must have struck many in the cost conscious British Rail as odd. Even if this plucky little group could raise the deposit, which seemed a stretch, was it credible that they could raise the full amount? There must have been those around the British Rail board table who doubted it.

In the meantime, with the belief that only real passion can bring, the Society members continued to work on the line as though the conclusion were already inevitable.

Every Sunday, regardless of the weather, they turned out to clear the tracks and make much needed repairs to the station buildings. Many of the volunteers came from far afield to spend their weekends working and would sleep in the booking hall or old coaches in Bridgnorth.

Finally, in February 1966 British Rail accepted the offer and agreed that the Severn Valley Railway Society could buy the line, with the deposit of £2500 now due on 14th May.

Sighs of relief were breathed all around before the Society moved on to their next challenge: finding the money.

Fundraising

Despite the huge sum that had to be found the Society went about raising funds in a way that is completely in keeping with the community project that the railway had already become. No sponsor stepped in to save the day with a large cheque, and no doubt that made the fundraising more difficult but it helped to establish the community feeling that runs through the railway to this day. No single person or organisation could claim a proprietary ownership of the SVR, a fact which would lead to a fierce fight a couple of years later (see Chapter Six).

The 'Save the Severn Valley Railway Fund' had been started in the public meeting in Kidderminster on 9th August the year before and the members' subscriptions of a guinea per year were paid into it, however at this stage the fund had not been open for a year and the number of members, while successful for so young an organisation, was still less than two hundred.

The first step was for the secretary, Bert Cleaver, to write to the existing members and ask for donations, but it was clear that the members alone would never be able to fund the line, so they set about a variety of other fundraising activities to raise money from the public.

Through 1966 a number of raffles, jumble sales, film shows and open weekends were run. Thousands of people attended the open days and took part in events like darts competitions and mini ten-pin bowling games. The only train rides available were provided by a 5 inch miniature railway set up on the platform that, despite its size proved to be a draw with the crowds!

Good progress was made throughout the year (for example by November the number of members had grown to more than four hundred) but despite their best efforts it wasn't possible to reach the full £2500 deposit by the deadline set by British Rail.

Thankfully British Rail stood by the agreement and on 23rd September the Society was able to pay an instalment of £1150 to cover the station buildings. The final £1400 was paid on 8th February 1967 and another milestone had been reached. Now, subject to contracts being worked out and a limited company being formed the Severn Valley Railway Society would be the owner of the railway.

The First Locomotives

Throughout 1966, in anticipation of the purchase of the line being completed, the Society had also been negotiating for locomotives to work on it.

On 25th March 1967, a dreary overcast day, they got their wish.

At lunchtime at Bridgnorth station both platforms, and the bridge above them, were filled with excited enthusiasts, all braving the cold and the showers while they waited for a special arrival.

In the background a fairground organ played popular tunes to keep everybody's spirits up in spite of the weather. At 12.15 the organ was playing 'Pack up your troubles in your old kit bag' when it was interrupted by a whistle coming down line from the direction of Bewdley.

A ripple ran through the crowd and a minute later they got what they had all been waiting for: no. 3205, an 0-6-0 Collett 2251 class, hauled a set of four GWR coaches around the bend in the line!

Approaching the station tender first, the engine set off a series of loud cracks from the detonators that had been placed along the track, and as it passed under the footbridge cutting a ceremonial tape that had been stretched underneath, the driver blew the whistle to the delight of everybody present.

A fantastic moment had arrived, the Severn Valley had a line and an engine and coaches to use on it!

***0-6-0 3205* was the first engine to arrive on the line on 25th March 1967.** *Here she is three years later at Hampton Loade on the official opening day, 23rd May 1970*

3205 was owned by the 2251 Fund started by David Rouse and would remain in use on the line for the next twenty years, starting the approach the railway has continued to take of focusing its ownership on the line itself and making arrangements for the engines of others to be used on it[1].

The coaches were all GWR stock; two non-corridor Churchwards, numbers 3755 and 3756, and two Collett composites, 7285 and 6562. The first three have subsequently been given to the Great Western Society at Didcot Railway Centre, but the last, 6562, is still in active service on the SVR.

The next three days were spent giving the assembled visitors rides up and down the two thirds of a mile between the station and Oldbury viaduct.

[1] Several engines have, of course, been owned by the Holding company over the years but they have always been the minority.

Oldbury Viaduct *above Daniel's Mill, just outside Bridgnorth station, this was as far as the first rides on the SVR would go. The engine in the shot is indistinct but appears to be a prairie, probably 5164 or 4566.*

Four weeks later, on 22nd April, LMS 2-6-0 Ivatt class 2 no. 46443 arrived having only been withdrawn from mainline service with British Rail the previous month.

LMS Ivatt class 2, 2-6-0 no. **46443 was the second engine to arrive on the SVR on 22nd April 1967***. Pictured here at Bridgnorth yard a year later on 17th September 1968. (0-6-0ST 2047 is just visible in background.)*

And here in 1970.

On her first day on the line 46443 made several runs to Oldbury Viaduct light engine, with Society members taking it in turns to ride on the footplate.

The engine would become a firm favourite on the line, racking up more miles than any other in the first decade of operations.

46443 adorns the sign of the Railwayman's Arms in Bridgnorth (the other side shows 7802 Bradley Manor).
46443 has long been a favourite on the line. In the mid-eighties a campaign to prevent it going to another railway was run with t-shirts printed bearing the slogan '46443 The People's Engine"!

The final engine to arrive in the Spring of '67 was GWR diesel railcar no. 22. On 13th May 1967 3205 towed it from Alveley Sidings to Bridgnorth.

No. 22's arrival actually marked a return for the railcar to the line. It had worked the Severn Valley in the days of British Railways between 1958 and 1961.

Diesel railcar W22W (no. 22) was the first full restoration on the SVR.

The Battle for a Light Railway Order

As we saw in Chapter One, all of the railways that exploded onto the scene in the Railway Mania of the mid to late 1800's had to have their own Act of Parliament passed to authorise them.

In 1896 that was changed by the introduction of the Light Railway Act. From then on to start a railway like the Severn Valley Railway Society was planning would not need a specific law to be passed by parliament, instead they would have to get what was known as a Light Railway Order signed by the Minister of Transport.

In theory this was much easier, but in practice for the Society to get their line it meant a two step process.

First, British Rail, as the current owner of the line, had to apply for an Order to downgrade it to a Light Railway. Once that was granted, a second Order had to be applied for to transfer ownership to a company the Society would set up.

0-6-0ST 2047 Manning Wardle came to the SVR on 22nd October 1967. *Owned by the Warwickshire Industrial Locomotive Preservation Group. 2047 was painted in blue as shown here, in the same month as her naming ceremony on 3rd April 1971. In 1995 she was repainted into her original working livery of green and stood outside the Kidderminster Railway Museum as a static display until 2010 when she was moved to Bridgnorth for repairs.*

It is a cumbersome process but easier than getting an Act of Parliament passed, and if it seems an odd way to go about preserving a railway, we should remember that was never the original idea. When the Light Railway Act was introduced nobody would have imagined that two thirds of the lines in the country would be closed within the century, much less that people would try to save them.

Since a Society was not legally allowed to operate under a Light Railway Order a new limited company was formed, the Severn Valley Railway Company Ltd, but for now this was a formality and the Society continued to handle all public matters.

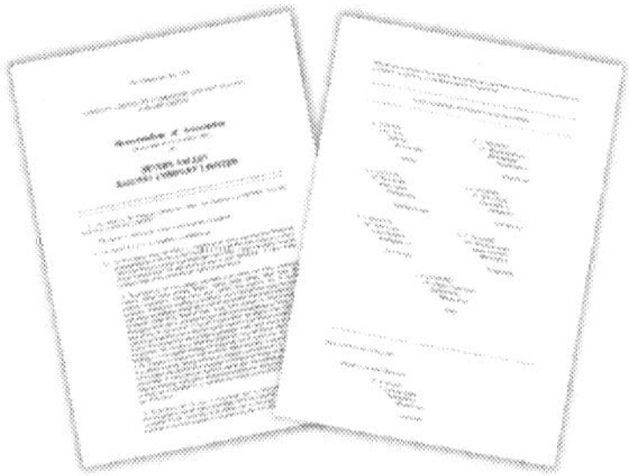

The Severn Valley Railway Company Limited certificate of incorporation*, showing the goals of the company and the original directors. This company, formed on 24th May 1967, is known as the* ***Guarantee Company*** *in contrast to the Holding Company we will meet later.*

On 1st February 1968 British Rail's London Midland Region announced that they were applying for a Light Railway Order. Although more complicated, the process is similar to planning permission in that people or organisations who might be impacted by the Order are allowed to object, but nobody was seriously anticipating any objections to be raised.

The biggest objections that the Society had faced up until this time were in its earliest days and they were from fellow preservationists who felt that the SVR was a threat to their own plans for other nearby lines. Now that the Society had made so much progress none of those groups would have formally objected.

0-6-0ST 813 arrived on 25th November 1967. *She was built in 1900 for the Port Talbot Railway which ultimately became a part of GWR. As of August 2016 she was back in operation on the SVR.*

0-6-0T 47383, the LMS Jinty arrived on 26th May 1968. *Seen here on the sidings at Highley with a freight train on 10th September 1983.*

However, British Rail and the Society were equally surprised when instead objections came from an unexpected source. Highley Parish Council objected to the Order on the basis that they planned to carry out roadworks around Eardington and reopening the railway would prevent them. That was bad enough but the biggest objection came when Shropshire County Council supported Highley and added their own objection.

According to their objection they were planning to build a by-pass around Bridgnorth and the planned route of the road ran across the railway just south of Bridgnorth station. Although the idea of the by-pass was not new (it had first been proposed as long ago as the 1930s) it hadn't been raised as a concern before this point as nothing seemed to be happening with it. Now though the council used it as the basis of a formal objection.

The Society members were horrified, the by-pass would either cut off the railway removing the station that formed the terminus (and their headquarters) or a costly bridge would have to be built.

British Rail, as the official sponsors of the Order, strenuously defended the Society, calling the objections "premature, irrelevant and inappropriate."

2-6-0 43106, the Flying Pig, joined the SVR on 2nd August 1968 under its own steam.

Negotiations with the council rumbled on through 1968. In the meantime new engines arrived on the line and fund raising efforts continued. Galas were run in April and on the late August bank holiday weekend. Because the Light Railway Order wasn't in place it wasn't possible to offer fare paying rides to the public, but members of the society were allowed, so a work around was found: nobody bought a ticket to ride, instead they bought a day membership of the society and then rode for 'free'.

Unfortunately it was not possible for the council, British Rail, and the Society to come to an agreement. To settle the matter the Minister called for a public enquiry to be made in October 1968.

More than a hundred Society members attended the enquiry to show their support, and British Rail's representative argued that,

> *"I fail to see why the company should be frustrated in their attempts to re-open the line simply on the chance that the by-pass might go through at a later date or never even go through at all."*

Immediately after the enquiry the Society put on a train pulled by 3205 to allow anybody concerned to see the planned line. The Inspecting Officer was obviously suitably impressed as his official report recommended that the Order be granted.

The Society members must have been relieved but their relief was short lived.

The process is a slow one but when the Minister finally commented on the Order in June 1969 he overruled his Inspector's recommendation and refused to grant the Order. He stated that if Shropshire County Council did choose to build the by-pass, then it would not be appropriate for the taxpayer to have to pay the extra cost to build a bridge because of the railway. However, instead of rejecting the Order, he gave the Society four more months to come to some agreement with the council.

In the end, after a three and a half hour meeting with the council, the Society found a way to get them to withdraw their objection. They agreed that if the council ever decided to build the by-pass, then the Society would provide SVR land for the road to run through, and either pay at its own cost for a bridge to be built, or sacrifice Bridgnorth station.

None of it was very palatable but it was the only way out of the stalemate, and it did mean that on the 4th December 1969 the Minister of Transport was finally able to grant the Light Railway Order to British Rail. Of course, that wasn't the last part the by-pass bridge had to play in our story, but the first legal step towards the re-opening of the line was now complete.

Two days later, to prepare for the LRO to be transferred to them, the Society formally merged with the limited company that they had formed three years earlier. Then, on 15th May 1970, the Minister granted the Transfer Order, which moved responsibility for operating the line to the Severn Valley Railway Company Ltd.

A week later, on 23rd May 1970, the line between Bridgnorth and Hampton Loade was re-opened to passengers.

At 14:00 the Mayor of Bridgnorth blew the whistle and 3205 left the station, pulling the first train of the new Severn Valley Railway!

0-6-0 3205 at Bridgnorth on official opening day 23rd May 1970. *The headboard reading 'Severn Valley Reopening Train' can just be made out beneath the boiler.*

Only five years after their first meeting, a small group of dedicated amateurs and volunteers had pulled off a monumental achievement, they had taken an abandoned branch line and resurrected it.

Whatever happened next at least the fight to save it was over and they could at last get down to running their railway.

Sadly, what happened next was that it was very nearly taken off them.

OPENING DAY, 23RD MAY 1970

2-6-0 43106 leaving Bridgnorth on opening day

2-6-0 46443 on opening day approaching a group of enthusiasts.

Also in operation on the opening day were Railcar 22 and 2-8-0 48773, running as LMS 8233.

CHAPTER SIX: A 20th century power struggle

When the line re-opened in May 1970 the Society owned the track from Bridgnorth to milepost 144½ at Alveley sidings as we have seen, although it only operated it as far as the Hampton Loade station. [2]

However, the previous year, while the fight for the Light Railway Order was still happening, Alveley colliery had closed and then, in January 1970, British Rail had stopped passenger services between Hartlebury, Bewdley and Kidderminster. Those two things together meant that the whole line between Alveley and Kidderminster was now available.

This presented the Society with a problem. Obviously it would be ideal to secure the line to Kidderminster and with it a connection to the main line, but at the time they were still raising funds for the purchase of the northern section of line. It must have seemed a distant hope to embark on

---

[2] Actually the Society didn't 'own' the line when it opened, the balance of payment hadn't been made to BR. That wouldn't have been a problem until it was discovered that the Light Railway Order only legally came into effect when the final payment had been made, so for three weeks in June operations had to stop again! A final fundraising effort (including loans of £4300 from two generous members) meant that the final amount was paid on 24th June and the line 're'-re-opened on 27th!

negotiations for the southern section, nonetheless that is exactly what they did.

0-6-0PT 1501 *arrived on 17th October 1970 having spent only twenty years in service (11 on BR and another 9 with the Coal Board).*

On 15th October 1970 a price of £74,000 (c. £969,000 in 2019) was agreed between British Rail Board and the Guarantee Company covering the line from Alveley to just shy of Foley Park Halt, leaving the company with the now familiar problem of raising more money.

Having spent years raising the money for the original stretch of line the volunteers must have found this daunting and, in a curious repeat of history, they found what briefly looked like the answer in the form of a famous, and wealthy, member of parliament.

ENTER THE SHOWMAN

Sir Gerald Nabarro was the preservation era's answer to Sir Samuel Morton Peto who we met in Chapter One.

Like Morton Peto he was a rich, flamboyant MP and shameless self-promoter who ultimately suffered a very public fall from grace but was nonetheless instrumental in getting to the line we have today.

Born in 1913 to a family of modest means, Nabarro left school at 14 and after a brief stint in an office, ran away to join the Merchant Navy. Three years later he signed up to the army and worked his way up to Sergeant. In 1936 he chose not to pursue a career as an officer because he couldn't afford to support himself and instead left and started work as a labourer in a sawmill factory. He quickly rose to become a manager and later started his own sawmill which made his first fortune.

In World War II he was formally part of the army again, now as an officer, but his work in industry meant that he did not see action abroad.

He had a passion for expensive cars and drove a fleet of nine cars including three Daimlers, all distinguished with the private number plates NAB 1 to NAB 10 (he missed out on NAB 9).

In 1950 he became the MP for Kidderminster and in 1963 he was knighted in the new year's honours.

When he first became involved in the Severn Valley in late 1970, he was one of the most well-known politicians in the country. His handlebar moustache was a very frequent sight on TV and the papers, and his booming voice (with an accent that was described as Terry-Thomas like) often graced the radiowaves on shows like *Any Questions?*

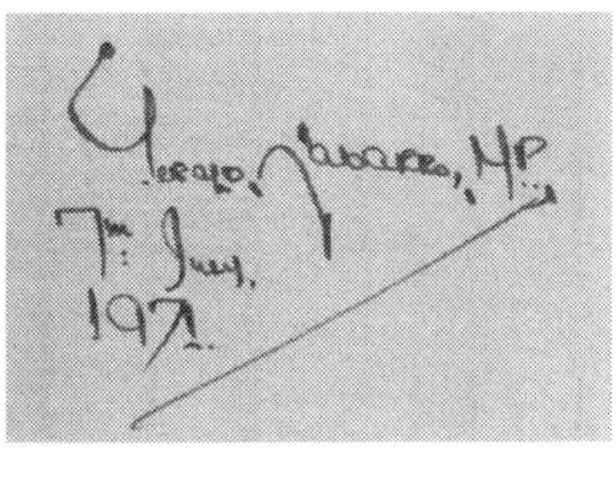

Sir Gerald Nabarro, MP. *The flamboyant, moustachioed MP who helped raise money for the line to Kidderminster but became involved in a very public fight for control of the SVR and an unconnected national scandal.*

Nabarro was a self-described railway enthusiast his whole life. Before getting involved with the Severn Valley he was the president of the Kidderminster Model Railway Club, so when the SVR opened in his old constituency, he naturally became interested.

Just like Morton Peto, Nabarro's original involvement must have seemed like a blessing. His showman like nature guaranteed publicity to the railway at a very early stage in its career when attention could only be a good thing. Unfortunately his fall from public grace overlapped almost exactly with his time with the SVR, whether it contributed to the difficulties with the SVR, or whether the two were just both symptoms of the same arrogance is for the reader to decide. Although his public problems aren't connected to the SVR, because they overlap so much in time and serve as an illustration of the character of the man we are dealing with, both stories will be told in parallel here.

4-6-0 Black Five 45110, arrived on 20th August 1970. *45110 was the last British Rail owned engine to pull a mainline steam train on 11 August 1968. Seen here in April 1971, 45110 was a stalwart performer on the SVR over two boiler tickets. She is currently on display in the Engine House.*

On 1st November 1970 Nabarro was made a director of the Guarantee Company.

At the time there were two huge financial threats looming over the railway.

The first was really an opportunity, i.e. the chance to purchase the line from Alveley to Kidderminster, but they did not have the £74,000 necessary to do it and letting it slip through their fingers must have felt like a terrible thing to miss.

The second was the threat from the Bridgnorth by-pass which we saw earlier. The agreement with Shropshire County Council meant that at any point the railway could be faced with a gigantic bill for the construction of a bridge completely outside of its control. Finding money was clearly a priority.

***4-6-2 Pacific 70000 Britannia pulled King George VI's funeral train** in 1952, before coming to the Severn Valley on 9th April 1971. She left the SVR in 1981. Seen here outside Highley in the 2018 Steam Gala.*

Nabarro suggested that the answer to both issues was to close the existing Guarantee Company and transfer its assets, i.e. the line the volunteers had worked so hard to save, into a new limited company.

The new company, of which Nabarro would coincidentally be chairman, would be called Severn Valley Railway (Holdings) Ltd and would raise the new money they needed from the public by issuing shares. The Guarantee Company members would be compensated by holding 36% of the shares in the Holding Company.

Nabarro spelt these ideas out at the Annual General Meeting of the Guarantee Company on 1st May 1971 but the meeting did not come to a decision.

Most sources date the breakdown of the relationship between Nabarro and the members to the next year, but the reluctance to reach a decision at the May AGM (and the subsequent forming of the SVRA as we shall see) indicates that at least some of the members were already becoming suspicious of Nabarro's motives even at this early stage.

A follow up meeting was set for 6th August.

THE FALL BEGINS

Three weeks later, unbeknownst to anyone on the SVR, the incident that started Nabarro's fall from favour with the general public happened, in Totton, Hampshire.

Nabarro and his company secretary, Margaret Mason, were in his Daimler (the one with the private plate NAB 1), when it was seen repeatedly weaving at speed in and out of a number of slower cars that were queuing to reach a roundabout, forcing two of them off the road. When it got to the roundabout the Daimler sped the wrong way around it, causing a lorry coming the other way to swerve to avoid hitting it.

The incident was reported to the police and they began an investigation.

Back on the Severn Valley the members' focus was on Nabarro's intentions for the line rather than his driving.

Although the Society had been wound up, on the 2nd July 1971 the former committee members got together and expressed their concerns about the direction of events. They supported the formation of the new Holding Company to raise the money to extend the line and prepare a defence for the by-pass bridge but were afraid that closing down the

Guarantee Company would concentrate too much power in the hands of Nabarro.

They agreed to suggest to the full membership that a new organisation, the Severn Valley Railway Association, was created to protect the interests of the members. The members agreed and John Garth was elected chairman of the Association.

Mr Garth, it might be remembered, had been involved form the very earliest days of the Society. He was the one to send the telegram that stopped BR removing the track at Bridgnorth and he had been chairman of the Society until it was merged with the Guarantee Company. Above all, he had the trust of the membership.

At the same time the chairman of the Guarantee Company, Andrew Marsden-Smedley, wrote to the members recommending that they accept the proposal to shut down the Guarantee Company and form the Holding Company.

He also said that to help the transition he would step down as chairman in favour of Nabarro.

On 6th August 1971 the delayed AGM was held and the members, no doubt mindful of the fact that the Association would do all it could to protect them, approved the formation of the new company. At the next board meeting in September Andrew Marsden-Smedley stood down as he had promised and Nabarro was made chairman of the Guarantee Company.

2-10-0 600 Gordon, arrived on 20th September 1971. *Shown here at Stockton & Darlington in 1975 for the Steam 150.*

By November the police had concluded their investigation into the incident at the roundabout and they charged Nabarro with dangerous driving. On the 24th January 1972 he was tried at Winchester Court and astonished everybody by pleading 'not guilty'.

He claimed that he had not been driving the car, that in fact his secretary, Margaret Mason, had been the driver. Mrs Mason backed him up, taking the stand and testifying that she had been driving at the time but didn't remember anything like the scene the witnesses were describing.

The jury didn't believe either of them and Nabarro was convicted of dangerous driving and fined £250 (about £2800 today). The judge was particularly scathing in his summing up:

> *"This was not an error of judgement. This was a deliberate and outrageous piece of driving..."*

The press, loving to tear people down even more than they enjoy building them up, went wild. Nabarro's huge moustache had always been a cartoonist's dream and when the verdict was announced the Evening Standard ran a cartoon of a police line-up of women, one of them complete with handlebar moustache!

Nabarro however, in a pattern that would be familiar to the members of the SVR, was unrepentant. He maintained his innocence throughout and said that he would appeal.

Six weeks later, on 15th March 1972, the Severn Valley Railway (Holding) company was legally formed with Sir Gerald Nabarro as chairman.

He was now chairman of both Severn Valley companies.

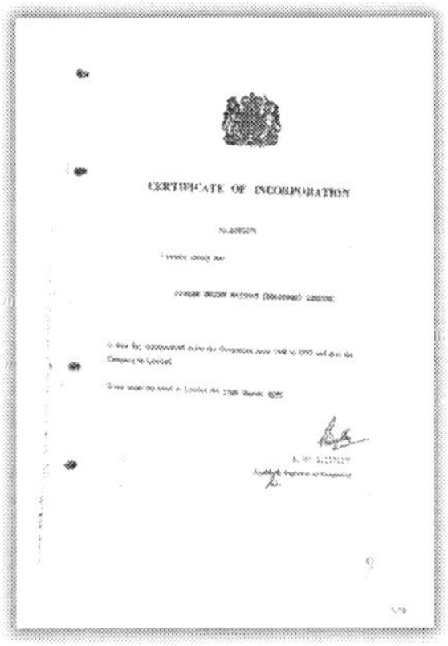
CERTIFICATE OF INCORPORATION

The Holding Company was incorporated on 15th March 1972.
Shares were at first only available to members of the Guarantee Company, but on 4th May they were released to the general public. By 22nd May the offer was closed and it had been oversubscribed.

Despite the increasing power struggle that was going on behind closed doors, and Nabarro's non-railway related difficulties, the idea of raising money via an open share offering was an ingenious solution and the public responded immediately, by 22nd May when the offer was closed it was oversubscribed.

The £110,000 had been raised and an extra £24,000 had to be returned to unlucky applicants.

2-6-0 61994 'The Great Marquess' *arrived on SVR 9th September 1972.*

In June 1972 Nabarro took his driving conviction, which of course was well known to the SVR members by this time, to the Court of Appeal and asked leave to appeal it. The Court, in what was at the time a unique answer for a driving offence, granted the MP his appeal.

At the appeal Nabarro was able to produce four new witnesses who said that they had seen a woman driving the car.

The prosecutor clearly mocking Nabarro's defence said,

> *"...but there were only two people in that car, the man, balding and with a great deal of hair on his face and the woman, blonde hair but not very long. Can you imagine anyone confusing these two?"*

He asked Nabarro to stand with his back to the jury and pointed out that his moustache was even visible from behind!

Cartoon appearing in the London Evening Standard following Sir Gerald Nabarro's acquittal.

Despite the obvious improbability, the jury acquitted him and Nabarro returned to the SVR an innocent man.

Just over a week later, on 3rd November 1972, he was at the Guarantee Company AGM to face an angry group of volunteers.

The undoubted success of the fundraising effort had done nothing to quiet the rumblings of discontent on the railway and at the AGM it boiled over into a heated debate.

Nabarro's restructure proposed that the Guarantee Company's 36% of shares in the Holding Company would not have any voting rights. Instead the Holding Company board (basically Nabarro himself) would be able to cast those votes. Furthermore, it gave the Holding Company the right to appoint half of the directors of the Guarantee Company and its Chairman.

These two things together clearly amounted to a removal of any direct power from the Guarantee Company, which would be reduced simply to operating the railway with the members doing as they were told.

The members were up in arms.

To tackle the issues the Association, under John Garth, put forward three resolutions:

1. To reject the restructuring and insist that a new contract be agreed between the companies and that it include the Association;
2. That the decision about the Bridgnorth by-pass bridge had to involve an Extraordinary General Meeting of both companies and the Association; and
3. That the railway could not take on any paid staff without first offering the position to the members.

The members voted unanimously in favour of all three of them, but it still didn't take away their concerns.

The last two resolutions betray separate issues with Nabarro that were starting to be raised by the members.

They were growing increasingly worried that the railway was being run commercially, and even worse, that it might become an 'old boys club' with jobs and appointments being handed out based on mutual self-interest and friendship, not in the interests of the railway.

Nabarro had done little to dissuade them of this fear. Over the time of his involvement there were several occasions when he acted as though the railway were his private plaything. During publicity for the share issue, for example, he was interviewed on Midlands television and, looking very self-satisfied, merely agreed that the railway was his:

Reg Harcourt [interviewing]: *You must be the envy of a lot of people. You've got your own private railway, Sir.*

Sir Gerald [nodding and smiling]: *I know.*

Perhaps this was just a matter of style, but now a series of events took place that appeared to confirm the members' worst suspicions of Nabarro.

THE ENGINE, THE DEVELOPER, AND THE SUPERINTENDENT

Although as we have seen there had been growing resentment of Nabarro's approach almost since the beginning of his involvement, there were probably three things that brought the situation to a head.

On 6th January 1973 4-6-0 4930 Hagley Hall arrived dead at Bewdley.

Since the very early days of the Society, the SVR had concentrated its funds on the physical assets of the railway such as the track, station buildings and signalling equipment. The engines that ran on the line were almost entirely owned by other groups (or in some cases individuals). They only used company funds for engines when there really wasn't any other alternative.

In early 1972 following that same approach Richard Dunn, the SVR's solicitor, had intended to buy Hagley Hall himself with a group of friends.

Dunn, who had been instrumental in getting the original Light Railway Order to operate between Bridgnorth and Hampton Loade, and then again in leading negotiations with BR to buy the line to Kidderminster, was originally close to Nabarro. It was Dunn who had shown the politician around the Severn Valley when he first visited after it opened in 1970.

Now though, Nabarro insisted that they buy Hagley Hall with company money and relations between the two men cooled considerably. [3]

---

[3] Nabarro later insisted in an interview to the Times that it was the other way around, that Dunn wanted to use SVR money for the engine and Nabarro tried to argue against it. We cannot know with complete certainty, however, we can see a pattern in the events involving Nabarro which is why I have used the

The purchase represented £6,000 worth of company money that the volunteers felt could have been better spent.

***4-6-0 4930 Hagley Hall arrived on 6th January 1973** and was the subject of a falling out between Nabarro and Richard Dunn, at that time the SVR's solicitor and later Chairman of the Guarantee Company. Dunn felt that 4930 should have been bought privately but the previous June Nabarro had used SVR funds to buy the engine.*

The Developer

At the beginning of January 1973 it was noticed that a well known property developer, Terence Willis, had been appointed as a new member of the board of the Holding Company and the next day he was recorded as having purchased an eighth of the company's shares.

Willis and Nabarro were introduced by their stockbroker (they used the same one) towards the end of 1972, and shortly afterwards the two men had a meeting to discuss the SVR at the Savoy in London. Originally Richard Dunn, who in addition to being the company solicitor was by this point the Deputy Chairman, was invited as well but three days before the meeting was told not to come so the meeting took place with just Willis and Nabarro. [4]

---

interpretation above. I suggest the reader makes a decision having read the whole of this section.

[4] Nabarro denied that Dunn was ever invited.

Willis claimed that at the meeting Nabarro told him that if the SVR failed then there would be considerable value left in the properties it owned. [5]

Willis then arranged to buy a significant block of shares (some from Nabarro himself) and was appointed to the board of directors.

His appointment to the board was technically in breach of the rules that directors would be approved by both the Guarantee Company and Holding Company, but that wasn't the main thing that bothered the members.

They were concerned that Nabarro had a secret plan to sell off parts of the railway to developers like Willis. And worse, the most obvious thing to sell first would be the spiritual home of the railway's rebirth, Bridgnorth station.

Given that Bridgnorth had so recently been the site of the members' triumphant reopening of the line, the idea of it being sold from under them understandably caused emotions to rise to the surface.

Feeding their fear were a number of changes that were happening about this time:

- Firstly the company headquarters were moved from Bridgnorth to Bewdley;
- In addition several developments were planned that would allow Eardington to operate as a terminus, including some that were carried out and the membership knew about, such as converting the siding into a run-round loop, and others that didn't get off the ground and were kept much quieter, such as enlarging the station and adding a carpark;
- And finally, Nabarro had the Bridgnorth site itself valued.

---

[5] Nabarro acknowledged the meeting took place but insisted he didn't tell Willis it was a property opportunity.

***2-6-0 7819 Hinton Manor** is now on display at the Engine House in Highley but arrived at the SVR on 6th January 1973 with 4930, 5164 & 4141.*

Many of these were probably perfectly sensible preparations in case the by-pass meant that Bridgnorth station had to be abandoned, but the membership generally felt that they all indicated an intention to use the by-pass as an excuse to sell off the Bridgnorth station site.

They pointed to the investment by Willis and argued that since the company was set up in such a way that there was no expectation of financial gain from dividends (all SVR dividends were, and still are, payable in free rail passes) he must have been expecting to make money from the sale of the station.

Both Nabarro and Willis denied this was the specific intention but Willis did frankly admit that he was 'not particularly a railway enthusiast'.

Whether there was anything sinister about it or not, the board were clearly preparing for the possibility of abandoning Bridgnorth in a way that the members did not approve.

However, in the absence of clear evidence that Nabarro was abusing his position, the belief of the members was not enough on its own to push things over the edge.

The Superintendent

That flash point came when Nabarro fired the Operating Superintendent without consultation with anyone else. [6]

[6] Once again, Nabarro blamed Dunn for this, saying in the Times, "Dunn is guilty of disgraceful misconduct; after urging on me the sacking of Becker he ran out and turned a somersault and claimed he had nothing to do with it."

The members demanded that the superintendent was given his job back but Nabarro refused.

The volunteer workforce then took matters into their own hands and threatened to strike unless Nabarro resigned.

2-6-0 78019 on the Great Central Railway. *78019 was resident on the SVR from March 1973 (the end of Nabarro's tenure as Chairman) up until 1998.*

Nabarro's fame (or perhaps notoriety, given his recent legal problems in the eyes of his detractors) now acted against him. The struggle became very public and made the national news. It was reported in the London Evening News on 12th March 1973 and a cartoon the next day showed a group of suited City commuters trying to attack Nabarro for disrupting their rail travel only for one of them to shout,

> *"Hold it men, he's right! It only involves four miles of private track in Shropshire."*

For the dedicated members of the Severn Valley though there was no 'only' about it. A resolution needed to be found.

On 17th March a joint meeting of the Boards of both companies was held and a solution was agreed.

Sir Gerald Nabarro resigned as Chairman of both the Holding and Guarantee Companies.

Even at this stage though Nabarro didn't believe he was out. He remained a director, and in the press release explaining his departure, he stated his intention to stand for re-election to the Board in the next AGM.

He obviously reconsidered his position because on 23rd May he resigned as a director as well. He still felt himself blameless though, despite the fact that in his resignation letter he acknowledged that his intention had always been for the Holding Company to control the Guarantee Company.

> *"I have, as you know, become increasingly unhappy with the way the affairs of the Severn Valley Railway are being conducted and I find myself quite unable to support the board's proposals for the company's future.*
>
> *I am particularly concerned that... will not preserve the effective control by the holding company over the guarantee company."*

By the time the June AGM rolled around, things had moved on. He did not stand for re-election and indeed he was scathing about the company in an article run by The Times on the same day, 29th June 1973.

He claimed sole credit for the profit made the year before and said,

> *"If it had been left to the volunteers the whole lot would have been spent in the tuck shop."*

The twentieth century power struggle was over. Nabarro was out and the line gradually got back to normal with better relations between the volunteers and the boards restored.

The way Nabarro's relationship with the railway ended was undoubtedly a shame but, like the trouble with Morton Peto in the previous century, his involvement was probably necessary to help get to the line we know and love today.

His fame helped garner the railway publicity which saw membership, fundraising, and public support grow over the essential early period and put the company in a strong position to buy the line into Kidderminster.

Ultimately the good he did the railway, and the problems he brought it, both seem to have sprung from the same place in his character. As he said in his own words,

> *"I have been driven remorselessly by ambition all my life... I do not lust for power... but the driving ambition is there all the time and accounts for much that I have done, for good or bad, in my life."*

Sadly there was no opportunity for a reconciliation with Nabarro, he never returned to the Severn valley and died at his home five months later on 18th November 1973.

CHAPTER SEVEN: Expanding the Line

The remainder of 1973 saw the railway settle down to much smoother running, work to prepare the line and stations beyond Hampton Loade continued and the future of Bridgnorth was secured.

4-6-0 75069 *arrived in scrap condition on 31st March 1973 from the Woodham Brothers scrapyard at Barry, the source of many of the steam engines that have survived into the preservation era.*

In June 1973 an Extraordinary General Meeting was held to discuss the potential sale of Bridgnorth station. The hard fought for right of the members to be included in the decision meant that the vote was overwhelmingly in favour of retaining the station and raising funds to pay for the by-pass bridge when it became necessary.

With the council not making any immediate moves to build the by-pass, attention now turned southwards to the rest of the line.

***4-6-0 42968 arrived on 14th December 1973** as the railway was preparing for the opening of the line to Bewdley.*

REACHING BEWDLEY

A Light Railway Transfer Order was granted on 29th March 1974 which gave permission to the SVR to run from Hampton Loade all the way to Stourport Road Bridge, just before the old Foley Park Halt and the sugar factory.

In practice the SVR only intended to run as far as Bewdley station. Extension beyond that point would have to wait.

The plan was to start running from Bridgnorth to Bewdley in mid-April of 1974, unfortunately however, an inspection discovered that the underbridge a hundred and sixty yards outside Highley was not suitable for traffic and would need to be fixed before services could operate beyond that point.

Rubery Owen Ltd agreed to build a replacement as quickly as possible but it wouldn't be ready and in place on time to meet the original opening deadline.

As a result the decision was made to open the line in stages, sticking to the original date for trains to Highley and pushing back the opening of the section between Highley and Bewdley by a month.

On 12th April 1974 the first service from Bridgnorth to Highley was pulled by Railcar 22, with 2-10-0 no. 600 Gordon pulling the first steam hauled train later in the day.

Five weeks later Railcar 22 was again the first engine to pull passengers over a new stretch of line. Rubery Owen had produced all of the metalwork for the new underbridge two days before the opening of the section of line to Highley, but of course the permanent way department needed time to properly install it and finish the rest of the track safely. The last of the installation work was finished on 21st April and, after passing an inspection, services were able to start between Bridgnorth and Bewdley on the 18th May.

The first steam engines to run over the new length of line were 2-6-0 43106 which ran up line to Bewdley and the honour of being the first steam from the new motive power depot Bewdley was taken by pannier tank 0-6-0PT 5764.

'The Flying Pig' 2-6-0 43106 pulled the first steam train from Bridgnorth into the newly re-opened Bewdley station *on the 18th May 1974. Railcar 22 had pulled the first train a couple of hours earlier.*

Running to Bewdley tripled the length of the railway from 4.5 miles to 12.75 miles, and at the same time added Highley, Arley and Bewdley stations to double the number of stops on the line.

As can be imagined, this represented a significant step up in operations needed to support the line. Not only did all of the additional track need maintaining by the permanent way team, but it also increased the complexity of that maintenance by introducing seventeen new bridges including the magnificent Victoria Bridge, Bewdley Tunnel and three new viaducts.

In addition the new stations needed maintenance and staff to run them. In terms of logistics this was probably as big a job as the original opening of the Bridgnorth to Hampton Loade line, but it passed off without a hitch, right down to the brass band that played on opening day at Arley.

0-6-0PT 5764 on Victoria Bridge. *The bridge was re-opened as part of the extension of the line on 18th May 1974 at which time 5764 worked the first train out of Bewdley. This shot of the same engine pulling a goods train, was taken some years later, at some point in the 1980's.*

On 25th May 1975, a year and a week after the extension to Bewdley, the loop at Arley was opened which enabled trains to pass each other and increased the number of services that could be run.

LMS Ivatt Class 2 2-6-0 46521 *shown here in May 1967 at Woodham Brothers scrapyard in Barry. She was rescued from scrap on 20th March 1971 and three years later in July 1974 she was the first of many SVR Barry restorations to re-enter service.*

A year after that **on 25th May 1975, 46521 and 2-6-0 43106 made SVR history again as the first trains to pass on the re-opened Arley loop.** *The photo shows her in restored condition at Bewdley in August 1976.*

4-6-0 6960 Raveningham Hall arrived in 1977. *An example of the modified Hall design that Hawksworth introduced soon after he took over from Collett as GWR Chief Mechanical Engineer. O. S. Nock says of the modifications, "Although in outward appearance it looked almost the same, nearly everything about it was new." Seen here at Arley in 1985.*

4-6-0 7812 Erlestoke Manor entered service on the SVR in September 1979. *Seen here with an SVR Golden Jubilee headboard in 2015.*

7812's sister engine, 7802 Bradley Manor, joined her at the end of November 1979 intended as a source of spares for 7812. *In the end the decision was made to return her to working service and after a decade of restoration she began work in 1993.*

BRIDGNORTH BY-PASS IS FINALLY BUILT

In 1982 the SVR finally received notice from Shropshire County Council that the long threatened by-pass would now be built, albeit at a much reduced width.

However, by now the railway had more than proven its value to the local economy and the council was a lot friendlier. Under a renegotiated deal the council no longer insisted that the SVR paid the full cost, instead asking for only 30%.

The last train ran across the line ran on 31st October 1982 and the following morning the track was lifted and excavation of the embankment began, severing Bridgnorth from the rest of the line exactly as had been feared nearly a decade earlier. This time though there was no possibility of abandoning the station!

The Bridgnorth by-pass under construction in 1982. *This photo was taken from the railway, at the approximate point where the railway bridge would be built months later.*

The bridge itself is made of two steel girders, 76 feet long and weighing 18 tons each. The metalwork was done by

Rubery Owen (who had earlier built the underbridge at Highley to enable the line to run to Bewdley).

The track was put back in place over the new bridge on 27th February 1983, a week before the new season started on 5th March. The line re-opened on time without any difficulties and everything appeared to be fine.

2-6-4 standard tank 80079 (resident on the SVR since May 1971) was one of the first steam engines over the new by-pass bridge, *when she returned on 28th February 1983 from the Welsh Marches Pullman mainline tour with 43106 and 4930 Hagley Hall. (Seen here at Bewdley, 1981).*

Unfortunately, in order to build the new by-pass, the council had been forced to lower the existing B4555 road that ran between the Severn and the railway track. Since the B4555 was just 100 feet from the line that meant increasing the height of the railway embankment.

On 30th April, just two months after opening for the 1983 season the embankment gave way leaving the rails suspended in mid-air.

On 30th April 1983 the embankment just south of the new by-pass bridge collapsed minutes after 2-10-0 600 Gordon passed over it. *The collapse was the result of the increased height of the embankment and the torrential rain during the build. (Gordon shown in August 1982 just outside Hampton Loade)*

Repair work started immediately but for two and a half weeks, services could only run from Bewdley to Hampton Loade while the council and the SVR worked together for fourteen hour days, seven days a week to get the line reinstated.

On 18th May the line re-opened and the saga of the by-pass bridge, that had threatened to kill the new SVR before it even began, led to the introduction of Nabarro and then been so tied up in the battle to oust him, was finally over.

REACHING KIDDERMINSTER

Despite the rest of the closures in the Beeching period, British Rail kept Kidderminster station open. The SVR had always had a goal to reach Kidderminster and since 1974, as we have seen, had owned the line up to Foley Park, however, tentative discussions about connecting to the British Rail station at Kidderminster had led to nothing as BR could not share a station with a preserved line. So, apart from occasional trips on Enthusiasts Days (the forerunners of the later galas) the line to Foley Park was not much used.

In 1978 there was a hope that a station could be built on the site of the old goods shed, which at the time the National Freight Company used, but negotiations fell through. For a time no other site was available and a home for the SVR in Kidderminster appeared in doubt.

On 14th May 1983 a solution arrived when BR closed Kidderminster goods yard, which provided enough space for a new SVR station to be built alongside the existing BR one (despite the threat of closure, the BR station itself remains open for passengers to this day).

Negotiations took place between the Severn Valley and British Rail, and the two parties agreed to the princely sum of £75,000 for the last mile of track from Foley Park to Kidderminster Junction, a thousand pounds more than had been paid a decade earlier for the 8.5 miles between Alveley and Foley Park. At this time, British Rail retained ownership of the land the goods yard stood on and leased it to the SVR.

Kidderminster Goods Yard in February 1984. *British Rail closed the goods yard in May 1983 and the SVR began negotiations to use the site for a new terminus.*

Following the success of the share issue for the extension to Bewdley the SVR boards, under the chairmanship of Bill Broadbent for the Holding Company and David C. Williams for the Guarantee Company, agreed to make another issue to raise the money to purchase the line and build a new station to be called Kidderminster Town, to distinguish it from the BR Kidderminster station next door.

Work began in May and proceeded at an astonishing rate. Track connecting to BR's line had to be lifted, ballast laid for the new track bed, the platform built, signalling equipment installed and, of course, the new track laid.

The new Kidderminster Town station platform being built in May 1984. Installation of the new track and platform was completed in just eleven weeks. *At the right of the picture, you can just see a Pickfords van. The National Freight Company still owned the goods shed that had been considered as the site of the SVR's station in 1978.*

On 30th July 1984 the new station opened, although without any station buildings yet, tickets were sold from an old coach which acted as a booking office.

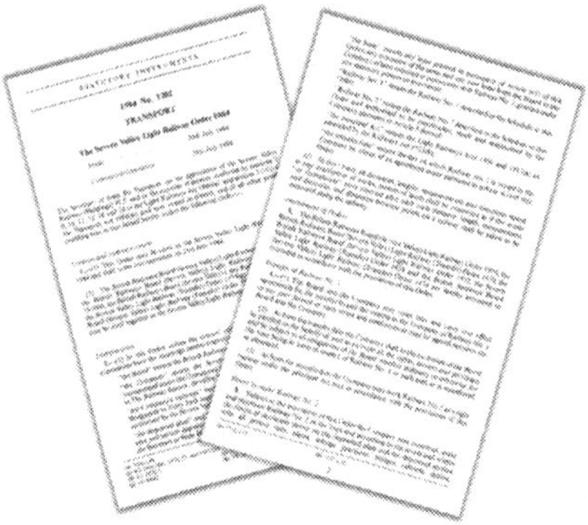

The Severn Valley Light Railway Order 1984 was approved on 20th July 1984 to come into effect the next day. *An inspection was carried out on 24th granting permission to open the line to Kidderminster.*

4930 Hagley Hall pulled the first VIP train out of the platform, and shortly afterwards 45690 Leander arrived from Bridgnorth, the first service to run the newly finished line.

The dream first spelled out in the Coopers Arms almost twenty years earlier had been accomplished and the line as we know it today was complete.

On 30th July 1984 4930 Hagley Hall pulled the first train from the new Kidderminster Town station. *Seen here on opening day.*

4-6-0 45690 Leander brought the first passengers up the line from Bridgnorth into the new station. *Shown here in LMS livery as 5690 two months after opening day.*

CHAPTER EIGHT: Travelling the Line

One of the most delightful features of the Severn Valley Railway is the beauty of the route it follows along the River Severn.

The line is blessed to run through arguably some of the most beautiful scenery in the country and that beauty is only increased by the stations themselves.

This chapter considers each station (and a couple of other points of interest) on the preserved line in the up direction from Bridgnorth. In this way we will replicate the order that the line was reopened, giving us the chance to consider some of the features that haven't been touched on in the history so far.

Naturally there is more to say about some of the stations than others, the largest section by far is the one that covers Kidderminster Town, where we will discuss the developments that have happened there since it was opened in 1984.

BRIDGNORTH

Bridgnorth is considered by many to be the spiritual home of the Severn Valley Railway. Despite the fact that the initial meeting of the group that would become the SVR took place in Kidderminster Bridgnorth was the site of the rebirth of the railway and it forms the northern terminus of the line to this day.

The station building is one of three Grade II listed buildings on the SVR, the others being Victoria Bridge and Wribbenhall Viaduct.

LOCOMOTIVE WORKS

Bridgnorth station has been the main home for locomotive works since the SVR reopened, and as of 2019 it consists of a loco shed, paint shop, machine shop and a boiler shop, however it was not always quite so well equipped as it is today.

When the SVR took ownership of Bridgnorth the only building on site apart from the station was a goods shed, which contained a small number of machines and the old loading platform.

Although the shed was used for whatever work could fit inside, most of the work on the engines had to take place outdoors which was clearly not sustainable.

TWO VIEWS OF BRIDGNORTH SHEDS FROM THE FOOTBRIDGE

September 1972 showing the old goods shed which would be converted to a machine shop in 1979.

The loco shed was built between March 1976 and April 1977, *the visible section at the front originally came from Portskewett in Wales. A year later the Heaton Mersey shed was added at the rear. (Photo taken in April 1984).*

A solution was found when a Locomotive Shed was sourced from Portskewett in Monmouthshire and transported to Bridgnorth. In March 1976 work started to erect it and ran through the year. By April 1977 the shed was in place, the track laid and illuminated inspection pits had been dug out to allow access to the underneath of the engines.

In 1978 an extension was provided by adding another loco shed behind the Portskewett one, this time from the Heaton Mersey diesel depot outside Stockport. Then gradually from 1979 onwards, as additional machinery was purchased, the original goods shed was converted for full use as a machine shop.

***The inside of the loco shed during an open day in 2018**, with 43106 displaying her open boiler.*

The loco shed and machine shop gave somewhere for engine work to be carried out, however, riveting of boilers was still carried out outdoors.

In 1986 Bridgnorth Council issued a noise abatement order to the SVR with instruction to stop outdoor riveting, this did eventually lead to the construction of the boiler shop, though not before a fair bit of nonsense!

As there was no indoor space where boiler work could be carried out there was no option for the SVR but to appeal against the order. This was done and the case, when it reached the Crown Court, was dismissed though only on the technical grounds that it was not Crown Court material.

This meant the first order was null and void and so the SVR could restart riveting, however, the council almost immediately issued another notice.

The SVR once again appealed and, to prevent it going to court again, the two sides agreed that permission would be granted for the SVR to carry out boiler work outdoors for two years, providing it only took place from 9am until 5pm during the week and that they raised a request for planning permission to build a boiler shop.

Boiler work resumed and, as agreed, a planning application was made for a new boiler shop, only to be refused by the council!

***Bridgnorth Loco Works in 2015 with the boiler shop** in the far distance. The old goods shed that now acts the machine shop can be seen sandwiched between the two newer buildings.* *Compare this view to the views on p. 131.*

Thankfully after some modification, a second planning application was granted in 1987 and after a share issue to raise the £500,000 needed, the boiler shop was built in 1989.

Inside Bridgnorth Loco Works – the boiler shop. *The boilers in the foreground belong to 75069 and 42968. The boiler closest to camera is an external contract for the Isle of Man (it appears to be 0-6-0T no. 15 'Caledonia')*

EXTENSION AND REFURBISHMENT

In recent years the SVR has embarked on the largest project Bridgnorth has seen since the line reopened.

Recognising that significant work was needed to update the facilities at Bridgnorth, a redevelopment project was kicked off and in 2012 a press release was issued with details of the long discussed refurbishment.

Originally the plans included a refreshment room, a turntable, a new car park, staff accommodation, a viewing area for the loco works, and refurbishment or conservation of the original station building. Over the intervening years the plans have changed, a number of the projects have shifted into later phases and some, like the ability for the public to view the loco works, have been dropped altogether.

However, over the winter of 2018 and 2019 the first elements were opened. The new carpark and toilet block opened in December 2018 and the refreshment room opened in January 2019.

The new refreshment room, Bridgnorth. *The building is in the style of a GWR station building from the early twentieth century. Planning permission was granted on 16th August 2016 and work started to clear the ground immediately. The building was subject to several delays but finally opened in January 2019.*

Designed by Robert Marrows (an SVR member who was also the architect of the Kidderminster Town station) and David Postle of the Kidderminster Railway Museum, the new building is styled after a GWR station building from c. 1910.

As at the time of writing, the later phases of the development including the locomotive turntable for which planning permission was granted on 30th April 2019, are still in development.

EARDINGTON

Eardington Station was first opened on 1st June 1868, six years after the rest of the line. It was built to serve Eardington forges rather than the village.

The forges had been opened by the Wheeler brothers, John and William, and made fine charcoal iron for gun barrels, horseshoe nails and wire. There are two forges, one on either side of the track and a level crossing three quarters of a mile down line of Eardington station joined the two together. Crossing Cottage, the house that sits alongside the crossing was once the home of the Eardington station master before the role was disbanded and taken over by the Highley station master in 1931.

4-4-0 3717 City of Truro on Eardington bank outside Crossing Cottage, *¾ of a mile down line of Eardington Halt.*

Eardington was run as a station through the GWR period and by British Rail until it was downgraded to a halt in 1952.

It closed with the rest of the line in 1963 and when the SVR reopened the railway in 1970 it formed the only intermediate station between Bridgnorth and Hampton Loade.

The station was closed again in 1972 when work began on the run around loop as part of preparing it for use as a northern terminus of the line should Bridgnorth be abandoned. Thankfully these plans came to nothing and Eardington reopened for a second time in 1981 and was included in timetables as a request stop until 1982 at which point it was abandoned and left to deteriorate.

In 1984 the platform was judged to be at risk of disintegrating and, to avoid debris ending up on the line, half of it was dismantled.

Eardington Station was removed from the SVR timetable in 1982 and left to deteriorate. *Seen here in 1994 after half of the platform had been taken out and replaced with the sloping concrete retaining wall on the left.*

It remained that way until a group of volunteers known as the Friends of Eardington Station formed and started a multi-year project to restore the station.

As can be seen by comparing the pictures below, they have succeeded in cosmetically restoring it, including notably through 2018 and 2019 reinstating the platform to a full 300 foot length.

The platform itself cost a total of £41,000, of which £13,000 was provided by the SVR Charitable Trust and the Board of the Guarantee Company, with the rest being raised by the Friends.

The restored Eardington Station in June 2018. *The difference in colour of the bricks marks the boundary of the old and new. Starting in November 2018 replica platform edging stones have been added as well.*

Amazingly, the bricks used to rebuild the platform are original GWR material having come from a GWR built bridge outside Bristol which had to be demolished due to electrification of the Great Western London to Bristol mainline. After being donated by the contractor carrying out the work all 6000 were cleaned by the Friends and then laid by a local building firm.

The edging stones were cast in concrete to match the original sandstone ones and were laid between November 2018 and February 2019 allowing Eardington Station to open for visitors (though not passengers) as part of the spring gala in March.

At the time of writing the station does not form part of the SVR services, however, the Friends are keen to get the station added back into the timetable for special events and potentially even as a regular halt at some point in the future.

HAMPTON LOADE

When the SVR reopened in May 1970 Hampton Loade was the southernmost terminus of the line.

Interestingly for reasons nobody seems to know, the station takes its name from the hamlet on the opposite bank of the river, not the village it is actually in.

Hampton Loade sits on the east bank of the Severn while the station is on the west in the village of Hampton. In the first timetable issued in February 1862 it was called simply 'Hampton' as would be expected, however, only a month later it was referred to as 'Hampton Loade' and that name has been used ever since.

The name 'Loade' provides a clue that this location has been used as a river crossing since at least Anglo-Saxon times, 'loade' comes from the Saxon word for 'ford'. There is no ford there now but a ferry has operated between the two villages since before the railway opened and until recently a current operated ferry, the last one in England, would take passengers across but at the time of writing it is no longer operational.

For the first twenty years after opening, Hampton Loade station only had a single platform and track on what is now the down line (i.e. heading to Bridgnorth) with a siding on the river side of the line.

In 1883 GWR added a second platform and a passing loop to allow trains to cross, albeit this is the shortest passing loop on the line with space for only a locomotive and eight carriages. A signal box was installed at the same time, unfortunately, however, much of it was demolished when the line was closed in 1963.

Hampton Loade station as seen in 2011. *The view up line towards Kidderminster has barely changed since GWR installed the passing loop in 1883.*

The signal box that stands today is a reproduction of the original containing parts from a number of prior GWR boxes, the remains of the original Hampton Loade box, Ledbury North End, Stourport and Pensnett.

The Paddock Railway 32mm garden model railway in the grounds of Hampton Loade.

Hampton Loade is also the home of the Paddock Railway, a 32mm outdoor model railway covering about 30 square yards, which runs on most Sundays when the SVR is open.

4-6-0 7802 Bradley Manor crosses the road bridge outside Hampton Loade.

COUNTRY PARK HALT

Country Park Halt is by far the newest of the stops along the SVR having been opened on 4th April 1996. The history of the halt is intimately connected to the Highley Mining Company.

The Highley Mining Company had been formed in 1877 when the Viggars and Scrivener families opened a colliery at Highley (as the name suggests!). They very quickly established a connection to the Severn Valley Railway as a means of distributing their coal.

The colliery was profitable for a long time and the miners continued to follow the seam of coal for many years while keeping only the two original shafts at Highley, however, the workings were taking them further towards Alveley and away from the shafts.

As a result in 1935 a new shaft was opened at Alveley and two years later the mines connected to the original tunnels from Highley. The company then transferred the men to Alveley Colliery and in 1940 they closed the Highley Colliery, leaving the shafts open for ventilation.

On 1st January 1947 the colliery was taken over by the National Coal Board as part of Clement Atlee's nationalisation plan which, just a year later, would create British Railways and transfer the Severn Valley away from GWR.

After the Highley Mining Company's Alveley colliery closed in 1969, the surrounding area was left derelict and covered in spoil heaps from the mine. In 1986 work started to reclaim the land and six years later, in 1992, the 126 acre Severn Valley Country Park was opened.

To provide further interest to the park Bridgnorth District Council agreed with the SVR Board to join the two attractions by adding a request stop for foot passengers (the halt can't be reached by car). It was sponsored by the council and funded by a grant from the European Commission.

The halt is served by a substantial brick shelter and access to the country park is by a bridge built in 2006 to replace the one that the Highley Mining Company built in 1937 (see the photo of its construction on the next page).

Alveley colliery. *The area covered by the colliery was abandoned after its closure in 1969 until 1992 when Severn Valley Country Park opened on the site.*

When Alveley colliery was operational a halt used to exist slightly up line of the current Country Park Halt to serve the miners, this was the Alveley Halt mentioned on p. 43.

Unfortunately when the SVR were preparing to extend the line to Highley in 1974 they discovered that the platform at Alveley Halt was collapsing and was in danger of falling across the line so it was demolished.

In 1996, when the current Country Park Halt was built, it was located on the site of the old sidings rather than the original halt.

The first Highley - Alveley bridge being built in 1936-37 *by Highley Mining Co. It has now been replaced by a 2006 bridge that links the halt to the Severn Valley Country Park.*

Country Park Halt with 0-6-0PT 5764

HIGHLEY

Highley station opened as part of the original line in 1862 and closed to passengers on 9th September 1963 at the same time as the rest of the line north of Bewdley. The SVR reopened it as part of the southward extension on 12th April 1974.

Due to the work on the underbridge described on pp. 116-117, Highley formed the southernmost terminus of the SVR from April to May 1974.

The station house and signal box are both original, the house dates from the opening of the line in 1862 while the signal box was installed in 1883.

The station house at Highley is the only stone built station house on the line.

4-6-2 34027 Taw Valley opposite Highley signal box. *The signal box is an 1883 McKenzie and Holland Type 3. Unlike most of the other signal boxes on the line it has stood largely unmodified since that time.*

Despite the series of tracks that can be seen at Highley, none of them are designed for passenger trains to pass. They are in fact two sidings which were added in 1883, at the same time as the signal box, to provide for the various mining connections from the Highley Mining Company. (It is possible for trains to pass here but only

0-4-2PT 1450 in the no. 2 siding at Highley. *Seen from the footbridge.*

if they are non-passenger carrying).

THE ENGINE HOUSE

With the growing number of engines resident on the SVR somewhere was needed to protect them from the weather while they were out of use and waiting to be overhauled or repaired. As a result plans were drawn up for a shed to be built on the site of the Highley Mining Company sidings which had been acquired by the SVR in August 2001.

At the time several sources of external funding were available and the decision was made not to make another share issue but rather to pursue funding from external bodies. As a result of that decision the goal of the project changed slightly. To secure money from many of these organisations it helps to have an educational purpose and so what had been envisioned as a purely practical shedding solution became a fully-fledged railway museum.

Construction started in March 2006 with the goal of opening in 2007 but due to the torrential rain in 2007 which led to the landslides described on p. 192 the opening was pushed back. When the line was able to reopen to Highley on 21st March 2008, the Engine House opened on the same day.

The completed Engine House from line side.

The project cost £5.5 million of which £3.5 million was provided by the Heritage Lottery Fund and just under £1 million by the European Union Regional Development Fund.

The line south of Highley leading to the Engine House. *The white fenced structure in the foreground is a replica cattle dock built by the SVR. Cattle docks were a common sight on rural lines when cattle were moved by train rather than by road, they consisted of a pen to hold the livestock and were usually higher than passenger platforms to allow the animals easy access to the wagons.*

In line with the original plan to make a storage shed for engines awaiting repair, rather than just a museum that engines go to at the end of their useful life, several locomotives have been on display in the Engine House and then returned to service later. They include, 1501, 7714, 4930, 7325, and 45110.

There have also been a number of visiting engines on display including no. 3 Sir Haydn from the Talyllyn Railway and 4-4-0 1000 which was on loan from the National Railway Museum between 2007 and 2011.

In addition to the engines themselves the Engine House always has at least one exhibition about railways and has been used to host several commercial events since its opening.

FOOTBRIDGE

In 1915 a footbridge was opened across the railway track linking the station platform with Highley village further up the hill. It stood until closure of the line but after that point was allowed to deteriorate.

When the SVR reopened the station in 1974 it was apparent that the bridge had become unsafe and needed to be demolished.

There was no bridge across the platform until 2009 after the Engine House had opened the year before and passenger volumes at Highley increased.

2-6-2T 5164 passing under the 2009 footbridge.

In November 2008 funding for the bridge was secured and, although it was delayed from the planned completion date of April 2009, it was finished that year and opened to the public on 21st October 2009 by HRH the Duke of Gloucester, the same day that he opened the Engine House.

ARLEY

Upper Arley village on the banks of the Severn.

J. Randall in his 1863 *Handbook to the Severn Valley Railway* considered Arley to be one of the most attractive villages along the line:

"...Its ferry on the river, its timbered cottages, partially concealed in green indentations of the hill, its grey church tower, and those of the castle near, are a picture of themselves; but when showers of blossoms crown the orchard trees in spring, or ruddy fruits hang ripe in autumn, the scene is more enchanting still."

The ferry has gone now, but the rest remains almost as Randall described it one hundred fifty years ago.

The ferry crossing of the Severn at Arley was first recorded in 1323 but it was closed in the 1970's having been replaced by the footbridge that now links the station to the village. The footbridge was officially opened on 22nd January 1972 and cost £43,500 to build (or approximately £526,000 in today's terms).

The footbridge at Arley was built in 1971 and officially opened a year later. *It replaced a ferry which had operated here since at least 1323.*

Arley station had been closed to passengers for almost a decade by the time the ferry closed as well, happily for us though the railway, unlike the ferry, did not stay closed.

The reopened SVR reached Arley in May 1974 but even as the footbridge was being built in 1971 work was already underway up at the station to prepare for its reopening.

Upper Arley ferry in 1924. *Note the absence of the footbridge that would be built almost fifty years later.*

Originally opened with only one platform, Arley underwent several changes in the first forty years of its existence. In 1883 as part of the general improvements to the line a passing loop was added with a second platform, at the same time the platform on the up line, next to the station buildings, was raised and extended. Twenty four years later both of the platforms were extended again.

After the line closed in 1963 the track in the up direction had been removed and the signal box demolished.

Relaying the up track to enable Arley to once again be a passing place was a key part of the new SVR's plans and in spring 1974, before Arley was even reopened, the Board had stated their intention to get the loop and signal box in place for the 1975 season.

The station buildings and platform. *The brickwork shows the evolution of the platform over the years, the bricks on the left are the newest, laid as part of the 1907 track extension; immediately to their right, two distinct layers can be seen, the lower level of stones show the original platform, the higher level of bricks was added in 1883.*

The permanent way team worked on the track while a replacement signal box was installed on the site of the original one. The box is an LNWR one that came from Yorton, while the lever frame inside it came from the original Kidderminster signal box (not the more recent one that the SVR built in 1987 which is discussed on p. 173).

The new passing loop was complete by 25th May 1975 and the signal box was commissioned just under a year later on 10th April 1976.

VICTORIA BRIDGE

It would be impossible to produce a history of the Severn Valley Railway without discussing the Victoria Bridge.

Designed by John Fowler, the bridge was built in 1861 as part of the original works carried out by Brassey, Peto and Betts, and has come to be symbolic of the line itself.

The metalwork is a mix of cast and wrought iron, both elements having been produced by the Coalbrookdale Company.

The Coalbrookdale Company had been founded by Abraham Darby in 1708 just outside Ironbridge, seventeen miles north-west of the Victoria Bridge. Darby was the first person to successfully smelt iron using a coke-fired blast furnace and, under his grandson Abraham Darby III the company produced the world famous bridge across the Severn at Ironbridge. By the 1860's when they took on the work for the Victoria Bridge, the Coalbrookdale Company was under the control of Abraham Darby IV.

The bridge has a span of two hundred feet and weighs approximately five hundred tons. When it was built it was the largest cast iron arch in the United Kingdom.

As an aside it's interesting to note that the design was so successful that Fowler built an almost identical bridge in 1864 for the Wenlock, Craven Arms and Lightmoor Extension Railway branch that left SVR at Buildwas. The second bridge was named the Albert Edward Bridge for

Queen Victoria's son (the future Edward VII). The main difference between the two bridges is that the Albert Edward has double track, while the Victoria Bridge only has a single track in common with most of the Severn Valley line as we saw in Chapter Two. Although as part of the compromise Fowler struck back in the 1860's the abutments are built to be able to take double track.

Victoria Bridge at Arley, *designed by John Fowler who went on to design the Forth Bridge and built by the Coalbrookdale Company who had previously built Ironbridge, Victoria Bridge was the largest single span cast iron bridge in the UK at the time it was built.*

Victoria Bridge has undergone extensive repairs at least three times in its lifetime, first by British Rail in 1954 when new track, sleepers and ballast were laid, and then by the SVR in 1979 and 2003.

The 1979 refurbishment included replacing the wooden decking with steel beams and repainting the metalwork. In 2003 it was repainted and, because some of the wooden sleepers that had been put down in 1979 had deteriorated badly and were retaining rainwater they were replaced again, this time with a steel mesh ballast deck which should prevent the problem recurring.

The central segments have inscriptions cast on both sides commemorating the people involved in the construction. *From left to right they read: MESSRS. BRASSEY & CO., CONTRACTORS; VICTORIA BRIDGE 1861 JOHN FOWLER, ENGINEER; and CAST AND ERECTED BY THE COALBROOKDALE COMPANY.*

In 1987 the Victoria Bridge was designated by Historic England as a Grade II* listed building, meaning it is a 'particularly important building of more than special interest' and in the top 8.5% of listed buildings.

NORTHWOOD HALT

As we saw in Chapter Three Northwood Halt was not original to the line, it was added by GWR on 17th June 1935 in an attempt to increase traffic.

In Northwood's case this was successful enough that a booking office was necessary, making it the only halt on the SVR where tickets could be purchased.

0-6-0 3205 passing Northwood Halt. *3205 was making a welcome return visit to the Severn Valley from its home on the South Devon Railway for the September 2010 gala.*

The buildings were simple, the booking office was a small corrugated iron hut with windows on three walls and a gently sloping roof, in reality little more than a shed, while the shelter was a standard, open fronted wooden design.

As is common with most halts (Country Park being unusual in this regard) there is a lane bringing traffic to the same area, meaning that a level crossing is necessary. At Northwood this has been the site of three accidents over the years, two took place in BR days and sadly involved fatalities, but the most recent was in November 2000 under the SVR.

On 25th November 2000 at 5:40 pm a Bedford van crossed the level crossing as 48773 was approaching with fifty passengers.

Luckily the engine had slowed to less than ten miles an hour and, despite the van being dragged along the line, nobody was injured although the driver was taken to Worcester's Ronkswood Hospital and treated for shock.

The Department of Transport quickly examined the crossing equipment and gave it a clean bill of health and the Railway Inspectorate accident report later confirmed that the SVR was not at fault.

The level crossing at Northwood Halt, *site of the accident in November 2000.*

None of the original buildings were left on site after closure so a new shelter and running in board were built by a local school in time for the reopening in 1974.

Since then both have been replaced. The current running in board dates from August 2008 as the previous one had reached the end of its life.

THE PAGODA SHELTER

Although the original shelter at Northwood, and indeed the 1970's replacement, were simple wooden affairs, the pagoda style shelter that stands on the site now is an authentic GWR style building.

GWR first introduced halts, which they originally spelt 'haltes' from the French where they got the idea of unmanned request stops, on 12th October 1903, just over a year before the SVR's first halt, Foley Park, was created. At first they had no standard shelter design but four years later in 1907 they introduced the pagoda (by which time they had also switched to the more recognisably English spelling 'halts'!)

For the next forty years GWR installed pagodas in many places across their network, focusing on branch lines in particular, although the only halt on the SVR to get one was Foley Park.

Given how practical and cheap they were, other railway companies used the pagodas as well, but it was the GWR who made them their own. They opened more than four hundred halts between 1903 and 1947, the year before nationalisation, many of which had pagoda shelters.

The shelter at Northwood was constructed by the Friends of Kidderminster Town Station using material from one of GWR's original corrugated iron suppliers, Joseph Ash Limited. It was erected in March 2006.

BEWDLEY

As well as being the junction where the Kidderminster Loop branched off from the main Severn Valley line, Bewdley had also served as the junction to the Wyre Forest line to Tenbury from 1864. Partly as a result of sitting on the two lines the closures of the 1960's affected Bewdley very differently to the rest of the Severn Valley Railway.

The closure of the Tenbury branch to passengers in August 1962, a full year before the Severn Valley Railway, led to a reduction in traffic through Bewdley. However, when passenger services stopped on the Severn Valley line, the Hartlebury - Bewdley - Kidderminster triangle was left open, granting Bewdley a reprieve from the September 1963 cut that impacted the rest of the SVR so brutally.

The reprieve was short lived though, as another blow fell in January 1964 when the Tenbury line was closed to goods and traffic through Bewdley reduced even further.

For the next five years in addition to the passenger services on the triangle, freight trains ran through Bewdley to serve Alveley colliery. That kept the station active, even if not quite as busy as in earlier times.

It was the closure of the colliery in 1969 that sounded the final death knell, and Bewdley (along with Hartlebury and Kidderminster) closed completely on 5th January 1970.

Thankfully, by that time the SVR was only five months away from reopening the Bridgnorth to Hampton Loade stretch and a future for Bewdley could be imagined.

On 18th May 1974 that future was realised when the SVR extended to Bewdley and the station became the southern terminus of the SVR for the next ten years.

Bewdley station buildings in May 2017.

VIADUCTS

Bewdley is the location of two of the line's viaducts, Sandbourne just south of the station and Wribbenhall immediately to the north.

Sandbourne Viaduct is three hundred and three feet long, has ten brick arches each with a span of 25 feet, which sit on sandstone pilings. It is interesting in that it doesn't cross a modern road or river, having been built to cross a depression in the ground which is lower than the line south of here and the station itself

Wribbenhall Viaduct is the third of the listed buildings on the line, being considered a Grade II structure by Historic England. Built mainly of red sandstone (which is perhaps more attractive than it is durable, given the amount of infill with engineering bricks that has been necessary over the years), the viaduct is one hundred and twelve yards long, and consists of seven arches, each with a twenty five foot span the same as Sandbourne Viaduct. However, there is also a single larger skew arch with a span of forty eight feet, nine inches that carries the railway over the Kidderminster Road.

4-6-2 34053 Sir Keith Park crosses Wribbenhall Viaduct, *just north of Bewdley station.*

UNIQUE FEATURES

Primarily due to its history as a junction station Bewdley has a number of features that are unique on the SVR:

Island Platform

Bewdley has the line's only island platform (although the plans for Kidderminster Town described the platform there as an island, it is really more of a peninsula!). The platform is accessed by a footbridge that dates from the late 1870's, the exact date of its build is uncertain however, records show that it was authorised in December 1877.

The canopy over the platform was originally built when the Kidderminster loop was opened in 1878 but it was much shorter than it is today. Some time between the 1903 and 1927 ordnance surveys the platform was extended to its current length.

In 2016 underwent a significant refurbishment using materials reclaimed from the old Snow Hill station in Birmingham which had been largely demolished in 1977 (a sign under the new awning at Bewdley refers to Snow Hill's demolition in 1978, it is possible that date refers to the area from which the canopy was retrieved).

The refurbishment was 'Highly Commended' by the National Heritage Railway Awards in December 2016.

The island platform at Bewdley *is a legacy of Bewdley's time as a junction of the SVR and the Wyre Forest Line. Here 4-6-2 60019 Bittern A4 Pacific stands at Platform 1. The canopy over Platform 2 and 3 was installed in 1878, extended in about 1900 and fully refurbished in 2016.*

Signalling

The fact that Bewdley was a junction meant that, alone of the stations on the SVR, it has two signal boxes, Bewdley North and Bewdley South.

Two boxes were first installed here on 24th October 1877, both have had significant changes made to their lever frames since that time (the North box in 1908 and the South several times) but it is unclear whether the buildings themselves are the originals. What is known is that they have not been changed in preservation.

Both are in use today operating the two different systems which control access over the north and south sections of the line.

The 'staff and ticket' signal system described earlier in Chapter Two, whilst an improvement over the previous system did not enforce a connection between the points and signals and so accidents along the rail system were still inevitable.

A driver ***passing a token*** *on entry to Bewdley north. The loop that can be seen is a carrier holding the token itself, a metal key shaped device that must be carried by the train to get access to the section of line.*

As a result a number of improvements were gradually made resulting in a system of interlocked signals and points with access to each section of line controlled by a token that was passed between pairs of stations.

This 'key token' system is the system in use from Bewdley North signal box to Bridgnorth on the SVR today.

A token is removed from the token instrument in the signal box at one station and delivered to the token instrument in the signal box at the next station. In this way the signalmen can be sure that the train has reached the next station and the line is clear for the next train to enter.

The section of track from Bewdley South signal box to Kidderminster on the other hand has been controlled by a system known as Acceptance Lever since November 1987.

The technical details are beyond the scope of this book (and its author!) but the basic idea is that the whole section of track has a small electrical current passed along it, making a huge electrical circuit joining two levers, one at Bewdley South and the other in the signal box at Kidderminster (described in the next section on p. 173).

It is therefore impossible for one to be opened without the other being closed and the signalman can be assured that only a single train is on the line.

The ***multiple tracks and sidings at Bewdley*** *from its junction history are put to good use by the SVR.*

2-6-2T 4566 has just passed Bewdley South box, while the Flying Pig waits on the left..

Bewdley MPD

In 1900 the 'Rock Siding' was created in the sandstone cutting to the north of Platform 3 and since the SVR took over they have used the area as to create a small outdoor Motive Power Depot (MPD), digging out a locomotive inspection pit by the winter of 1974.

The MPD allows for light repairs to be carried out without transfer to the full MPD facility at Bridgnorth.

A new inspection pit was made in the location of the original in early 2002 to enable the workers to stand upright underneath the engines.

34053 Sir Keith Park stands over **the loco inspection pit** *in the Rock Siding.*

The water towers in the picture above are not original to the site but were installed by the SVR before the line to

Bewdley was reopened. The one on the left, beside the inspection pit, is an LNWR tower from Hadley, while the one on the right (which stands at the end of the island platform) is a GWR example from Stow Heath yard in Bilston.

The only water on the line that is treated is at Bridgnorth and Kidderminster so these towers, in common with the one at Highley, are only used when strictly necessary on galas.

Bewdley Carriage and Wagon Departments

Both the Bewdley arm of the Carriage Repair Works and the Bewdley Wagon Department are based in the Down Yard.

Bewdley Carriage Department works in conjunction with the North Star works at Kidderminster but specialize in heavier bodywork repairs including restoring carriages from derelict condition. They are based in the Paint Shop.

The *Wagon Department* is based in the Goods shed and restores goods wagons for use in the demonstration goods train which is run on gala weekends.

2-8-0 2857 pulls the demonstration goods train towards Arley. The train consists of loose coupled wagons which are restored by the Bewdley Wagon Department.

KIDDERMINSTER TOWN

Kidderminster Town is now the starting point for around eighty percent of passengers on the SVR. It is also the most significant recent development of the SVR, as a result this section is the longest in this chapter.

Building a new station

These days Kidderminster is home to two railway stations, however, this wasn't always the case.

The original station was built as part of the Oxford, Worcester and Wolverhampton railway in 1852. When it first opened, a small wooden building served as both booking office and waiting room, however, it was not really big enough to handle the traffic and in 1859 a slightly more substantial building took its place.

The community were not noticeably more impressed with this building either and by early 1863 the Town Council were planning to demand that the West Midland Railway (who at this time ran the line, though it would be taken over by GWR in the summer of the same year) replace it.

Nobody was therefore particularly disappointed when, in the early hours of Sunday 15th February 1863, the second building burnt down.

Given the longstanding complaints about the original, a replacement building had in fact already been under preparation by the West Midland Railway at their depot in Worcester and they wasted no time in getting it put up.

The new station was an Elizabethan style, two story building housing the booking office, waiting rooms, refreshments rooms and the station master's house all in one building. When the Severn Valley opened the Kidderminster Loop in June 1878, this was the station to which it connected.

The Elizabethan style BR station building at Kidderminster which stood from 1863 to until February 1968. *The photo here was taken in 1963, one hundred years after it opened.*

The rather duller replacement BR building which has served mainline Kidderminster passengers since 1974.

In February 1968 the 'Elizabethan' building was pulled down due to dry rot and replaced in 1974 by the dull, utilitarian building shown on the previous page.

Thankfully visitors joining the Severn Valley at Kidderminster do not have to be content with this unprepossessing building as the beginning of their journey! As we saw in Chapter Seven, when the SVR reached Kidderminster in 1984 it was not possible for a heritage railway to share a station with British Rail and so the SVR opened their new station on the site of the old goods yard.

The (much more impressive) SVR Kidderminster Town station *based on the Ross-on-Wye GWR station was built in stages between 1984 and 2006. The roof crestings were copied from a discovery in a Ross-on-Wye scrapyard.*

The new station could not be called Kidderminster station because, of course, that was already taken by the BR station next door. So the name Kidderminster Town was chosen following the GWR practice of adding the word *Town* to whichever station was closest to the town centre. Despite being practically on top of each other, the SVR building just barely wins that distinction!

Of course, if the terminus was to be a success then the new station needed buildings.

Robert Marrows, an SVR member and professional architect, produced a plan based on a GWR standard design developed in the 1880's. Although the design had not been used in many GWR stations a fine example had stood at Ross-on-Wye until it was demolished in the late 1960's and there were photographs that could be used as reference material.

The design is in keeping with the GWR heritage of the line and is substantial enough to handle the traffic that comes from being a terminus. It also lent itself well to being built in stages, which is precisely what happened.

Building work began in Autumn 1984 after the line opened in July and the west wing, on the right in the picture above, was opened on the 24th November 1984 housing the station shop and toilet block.

The frontage containing the main entrance and ticket office, was completed slightly less than a year later on 28th Sept 1985, although the canopy over the entrance was added by the Friends of Kidderminster Town Station in 1996.

When the station first opened there was only one platform in use. ***On 25th November 1990 Platform 2 was opened allowing two trains to be in steam at the same time.*** *Here pannier tank 1501 and 2-6-0 46443 both wait in steam.*

In 1994 the land that the station stood on was bought from British Rail, giving the SVR full ownership of the line from Bridgnorth to Kidderminster.

Finally (or finally at the time of writing, at least!) the east wing and the concourse roof were added in 2006.

The original west wing and the concourse roof which was added in 2006, showing the impressive steelwork.

In 2015 the BR station next door (seen on p169) was scheduled to be demolished and replaced. In an ironic twist of fate the proposed replacement was a 'heritage' building in keeping with the SVR station! However, when the plans were revealed in 2018 the new station was a hyper modern glass building.

Safari Park

On 16th April 1973 the West Midlands Safari Park opened just beyond Bewdley station, though of course the SVR did not run past it at the time. It was not until 1974 that the SVR reached Bewdley but since that time there has been talk of making some sort of connection between the SVR and the Safari Park.

Originally a footbridge from Bewdley station was suggested but in later years, after the line to Kidderminster had been opened, discussions have centred on a halt being added.

Given the Safari Park's reliance on visitors having a car, a halt is unlikely to be of much use without some sort of accommodation that would introduce foot passengers, but there have been several times where it has been considered.

In 2011 planning permission was granted for a large hotel development at the Safari Park. The plans included a halt on the Severn Valley, but the development never happened. At the time of writing in 2019 the Safari Park is once again exploring adding accommodation and talk has resurfaced about a halt being added to service them, although in the author's opinion the small number of luxury lodges proposed means it is no more likely to happen than before, still it would be an interesting development and the link between two of the area's most important tourist magnets could only be a good thing.

Signal Box

Kidderminster signal box was built to an 1896 GWR design. *Here 45690 Leander steams past, light engine.*

When it was first opened Kidderminster did not have a signal box and only one engine could be in the station at any one time. In July 1986 work started on the substantial brick signal box which now sits just outside the station before the footbridge. The box, with its hipped roof and 'three over two' windowpanes, is modelled after a design introduced by GWR in 1896 and used until 1927.

Building finished in December 1986 and through 1987 the signalling equipment was installed, including four miles of wires! The result is a 62 lever frame that, following a couple of changes since that time, now controls the up and down line to Bewdley South (see the section on Bewdley on p. 160), both platforms, access to the carriage works and the connection to the main line. The signal gantry was the first one ever to be commissioned on a heritage railway.

Kidderminster signal box was commissioned on 21st November 1987 and contains 62 levers in a 22 foot long frame.

SHEDS AND WORKS

Not just the main embarkation point for passengers, Kidderminster has also become home to a number of important functions over the years.

North Star Carriage Works

Pickfords put the old British Rail goods shed up for sale in 1985 and the SVR purchased it for use as a carriage repair shop.

Converting it to a workshop to allow light repairs to carriages began immediately and was given a boost by including it in the 1986 SVRA Raffle with the goal of purchasing lifting jacks to service carriage bogies. By February 1988 tracks had been laid and the jacks installed and the shop became operational.

7812 Erlestoke Manor stands at Platform 1 in front of the North Star carriage works.

In addition to the jacks, it now contains two 19th century dynamo test rigs and dedicated machine, joinery, and paint shops.

One of the dynamo rigs in the carriage works.
Electrical power in carriages is provided by batteries which are charged as they travel by dynamos. These rigs are used to test the dynamos. The bank of bulbs at the top simulate the bulbs in the carriages, and amazingly they were issued by the LMS in 1941 and still work (they are also blue to comply with the blackout requirements).

Carriage Shed

Storing rolling stock in the open obviously exposes them to the elements, increasing the amount of maintenance that is needed, so the Severn Valley Railway has always had an intention to have a dedicated carriage storage shed.

In 1997 with the first phase of significant work at Kidderminster complete, attention turned once again to providing some weather protection for the growing body of rolling stock. An application was made to the Heritage Lottery Fund to support with the heavy costs.

The original plans involved a planned locomotive display building similar to the Engine House now at Highley. However, the initial request had to be amended significantly before the Lottery accepted it two years later, and one of the things to go was the locomotive display.

In 1999 everything started moving, on the 12th May the Lottery granted the funding and on 22nd June planning permission followed. By September work had started on the site with clearance being undertaken by the SVR staff, while the actual construction was contracted out.

In April of 2000 the shed was complete and came into operation, and the contractors left site.

The carriage shed on the right of picture. At almost a fifth of a mile long it is one of the biggest buildings in heritage railways.
The engine which is obscured by the gantry is the new build Peppercorn A1, 4-6-2 60163 Tornado.

The shed can store up to 62 carriages, is more than a fifth of a mile long and has four internal lines, plus an external one on either side of the building.

In 2006 a carriage washer was added to 'Road 6', the external line furthest from the SVR working, hidden from sight of passengers on the line.

Roads 2 and 3 inside the carriage shed. *Road 2 (on the right) has a pit to allow maintenance. Roads 4 and 5, on the other side of the wall on the left of the picture have a platform running the first half of the shed to make it easier for the volunteers to clean the carriages.*

Diesel Depot

More recently, in February 2015, a diesel depot was also added providing similar facilities for the diesel fleet that is resident on the SVR.

Plans were made and approved almost a decade earlier but the floods of 2007 meant that effort was diverted onto the essential restoration of the line.

On 30th January 2016 the diesel depot opened. It has three roads, inspection pit and lifting jacks, and houses a Mk 3a Convertible Sleeper coach which provides accommodation to staff when needed.

Continuing the proud tradition of award winning on the Severn Valley, the diesel depot won the 2016 Heritage Railway Association Modern Traction Award for its 'major contribution to modern traction preservation'.

CHAPTER NINE: Running the Line

The story so far has focussed primarily on the building and reopening of the line, the engines that use it and the development of the infrastructure that supports them, but of course there is much more to running a successful heritage railway than that.

This chapter brings the story up to date by looking at both some of the challenges that the railway has faced in recent years, and the commercial services which have become so much a part of running the line.

THE EARLY DAYS: FILMS, GALAS AND SANTAS

Early on the SVR recognised that the line had the potential to attract a variety of customers for different reasons and has always attempted to cater for as many as possible.

The three events in this section are a good example of how the railway has tried to fulfil its different needs over the years: renting out use of the railway for film and TV is driven almost entirely by the need to generate an income; the galas aim to satisfy pure rail enthusiasts; while the Christmas services help to broaden the appeal of the railway to a group that wouldn't otherwise necessarily come, generating an income but also planting seeds of interest that hopefully grow through the year in repeat visits.

Film and TV

Film was one of the earliest opportunities that was spotted to bring in additional income for the railway.

Not counting documentaries about the SVR itself or railways more generally, the line has been used for filming on forty eight separate occasions since it opened.

The first time the railway allowed camera crews to use the line as a location was in 1972 for an episode of the ITV comedy series *Doctor in Charge.*

The various films and TV episodes that have used the line as a location over the years have resulted in very different amount of time in the spotlight for the railway, running all of the way from the 2011 *Sherlock Holmes: A Game of Shadows* movie which lasts thirty seconds at most[7], through to the two year run of *Oh, Doctor Beeching!* which was filmed almost exclusively at Arley.

Filming has dropped slightly since the eighties and nineties but it is still a useful source of income.

The 2011 Sherlock Holmes film contained a scene in which **a woman is thrown from Victoria Bridge,** *the actors and train were not present but added later using computers. The bridge was real though!*

The most recent film to be released which was shot on the line was the 2016 *The Black Prince* about Duleep Singh, the last Maharaja of the Sikh Empire, however, film crews once again took over sections of the railway in July 2019 under heavy secrecy, prompting a flurry of media excitement about what was being filmed there, but SVR management remained tight lipped.

---

[7] Oddly the only live filming was of Victoria Bridge, the actors and even the train were added later with special effects. Additional rail scenes were shot at Didcot.

Galas

Galas, with more frequent services and a higher number of engines in operation than usual, have been held since 1975, although in those days they were called "Enthusiasts' Days".

On 10th April 1976 a spring gala was added and since then the steam galas take place in Spring and Autumn each year, with the Autumn Steam Gala being the larger of the two, and the only one to include nighttime running.

Between 1977 and 1984, before the line reached Kidderminster Enthusiasts' Days were the only opportunity for non-staff to travel over the SVR owned metals to Foley Park.

The original Enthusiasts Days used resident locomotives, the attraction coming from the sheer scale of the operation, and of course the fact that in those days many of the engines were only just coming to the line or completing restorations.

4-6-2 60103 Flying Scotsman visited in September 2016 for the Pacific Power event in place of that year's Autumn Gala. *The engine had visited the SVR previously in 1990, '91 and '94.*

With the passing of time visiting engines started to be introduced. The first of these, 4-6-0 7029 Clun Castle took part in the GWR Weekend in Spring 1982, the first Autumn visitor was 5051 Drysllwyn Castle, in 1984.

From 1990 all Steam Galas involved visiting engines alongside residents, with some very notable visitors including: [8]

2-10-0 9F 92220 Evening Star in 1983, the last steam loco built by British Rail;

4-6-2 60103 Flying Scotsman in 1990, then in green livery as LNER 4472;

4-6-0 6024 King Edward I in 1994, before the line was upgraded in preservation King class engines were prohibited from travelling on the line due to weight restrictions

4-6-2 60163 Tornado in 2009, a 'mini gala' was held in 2009 with Tornado, the first new build steam loco since Evening Star.

... and many others.

4-6-0 6024 King Edward I outside Highley.

Seen here on 10th June 2008 when it pulled the Royal Train with Prince Charles aboard. Prince Charles

[8] The date given here is the first appearance on the SVR, several of these engines made repeat visits.

Since at least 1986 Diesel Galas or Festivals have also been held, usually twice a year (although some years only had one and in 2006 it was missed altogether due to declining numbers).

A number of special Diesel events have been put on in recent years, recognising the anniversaries which have been arising.

In 2004 the SVR staged the Railcar50 event, supported by funding of £49,100 from the Heritage Lottery Fund, bringing together the largest group of first generation Diesel Multiple Units (DMU) in preservation. The event celebrated the fiftieth anniversary of the DMU by British Rail in 1954.

Sixteen railcars attended and, although four were not operational, the twelve that were in working order formed the longest DMU to be worked in preservation.

Unfortunately the SVR's former resident railcar W22W was not able to attend due to the cost.

In October 2018 a similar event was held in honour of the fiftieth anniversary of the introduction of the Class 50 diesels with ten working locomotives taking part. More than three and a half thousand passengers rode the engines, exceeding the target by more than 50% and proving the popularity of this type of event.

Santa Specials

The Santa Specials are a key part of the commercial success of the railway. Including the services run between Boxing Day and New Year's Day, they account for almost a fifth of the tickets sold each year.

The original idea to bring Santa (okay, his helpers!) onto the railway was borrowed from the Keighley and Worth Valley Railway who had successfully done so previously.

The first service between Bridgnorth and Hampton Loade was run by WD193 in 1973 and Santa has visited the railway every year since, even arriving in 1982 by helicopter!

Once the line had opened to Bewdley in 1975 Santa took up residence in his 'grotto' at Arley, and services ran from Bewdley to Arley for the next nine years with only one exception.[9]

Arley station with the remains of Christmas not yet fully removed *as 4-6-0 45110 gets ready to leave. The 'grotto' made of three ex-Royal Mail coaches can just be seen through the steam beyond the station building, the Black Five still wears a wreath and a lone snowman has escaped removal on platform 2.*

The grotto was made up of three decorated ex-Royal Mail carriages where the children would visit Santa and be given a present. Volunteers acting the part of elves guided the children and their parents (and in some cases grandparents!) to a specific Santa with the others being kept out of sight so that any child only ever saw one and the magic was kept intact.

When Kidderminster Town Station opened in 1984 Santa stayed at Arley, but the embarkation point for passengers switched from Bewdley to Kidderminster. Arley then remained the home of Santa until 2019, except in 2007 when the floods described on p. 192 meant that the line to Arley was closed so the service ran to Bewdley instead.

Kidderminster and Arley stations are both decorated and various local musical groups, choirs and brass bands,

[9] The exception being 1979, when Victoria Bridge was under repair and services had to revert to the original Bridgnorth to Hampton Loade.

perform for the guests who can buy seasonal goodies such as mulled wine or roast chestnuts from the various stalls that operate on the platforms.

Organising the service was a serious logistical challenge: present buying began in March, the music had to be booked by May, and Arley station had to be prepared from about October onwards. And that's before the services started.

Running the services took approximately one hundred volunteers to manage the stations and staff the trains that ran every half an hour.

In 2019 the services have been changed. Santa did not wait in his grotto at Arley for the children to visit him, but instead travelled on the train giving the children their presents on the carriages. Trains ran from both Kidderminster and Bewdley to Highley and were top and tailed. It will be interesting to see if the new format is as popular in the coming years.[10]

THE 1990's: THOMAS AND 1940's WEEKENDS

The decade after the line reached Kidderminster saw the introduction of these two very popular commercial events.

Thomas the Tank Engine

The SVR has had a long association with Thomas the Tank Engine, the Rev. W. Awdry, author of the Thomas books, had been an SVR member since 1980.

In June 1992 the SVR ran their first Thomas the Tank Engine weekend. The engine playing the part of Thomas was the East Lancashire Railway's 0-6-0T No. 32 'Gothenburg', a Hudswell Clarke engine that had been used by the Manchester Ship Canal (MSC) before preservation.

---

[10] If the author will be forgiven a personal interjection here, he spent more than ten years taking his three children on the Santa Specials so feels a certain reluctance to see them go, but such is the way of progress! Hopefully ten years from now another family will feel the same way about the new service.

The weekend was a roaring success with seventeen thousand visitors attending.

At the time the SVR's resident MSC tank engine, 0-6-0 686 'The Lady Armaghdale', was out of ticket awaiting an overhaul. The engine was (and still is) owned by the Warwickshire Industrial Loco Preservation Group, who had originally been formed to preserve 0-6-0ST 2047 'Warwickshire'.

The success of the Thomas weekend prompted the SVR to agree with WILPG to overhaul 686 to run as Thomas. Two years later the overhaul was complete and The Lady Armaghdale began life as Thomas.

Her first appearance as the little blue engine was at Didcot in June 1994, but by September of that year she was back at home starring the Severn Valley's own Thomas weekend.

The next five years saw 686 travel the country visiting other heritage railways to perform as Thomas, making her the second most travelled SVR engine with 43 appearances on other lines (behind 813 who has visited 67!).

The SVR's Jinty 47383 took on the Thomas mantle in October 2000, while 686 was undergoing an overhaul to address recurring boiler issues, but by 2002 686 was back in the saddle.

In October of 2002, 686 made two firsts by travelling to ZLSM in Simpelveld, Holland. It was the first continental Thomas show but also the first time an SVR engine had visited an overseas railway.

Seven years later 686 had to be withdrawn from service having suffered from a recurrence of her earlier boiler troubles and has not yet returned to operations. She has spent most of the time since then in the Engine House at Highley.

Escalating costs and the additional requirements of the company who license the Thomas brand, meant that the last Thomas event on the SVR took place in 2012, when the

Midland Railway Centre's Jinty 47327 played the role of Thomas.

THE THREE LIVES OF LADY ARMAGHDALE

In April 1971, shortly after arriving in 1969…

… as Thomas in May 2002…

… and back in red livery as Lady Armaghdale in the Engine House at Highley in 2011.

1940s weekends

In June 1994 the first 1940s weekend was held. The event has evolved over the course of time and now includes a variety of displays and live entertainment at stations along the line.

Over the years there have been flypasts from the RAF's Battle of Britain Memorial Flight team in their Spitfires and Hurricane fighter planes, and Dakota and Lancaster bombers.

Most stations have displays of vintage vehicles and the event is often used by re-enactment groups, adding to the 1940s feel of the line. Actors dressed in period clothing (and sometimes playing familiar roles such as Sir Winston Churchill and King George VI) interact with the passengers.

48773 as WD 307 at Hampton Loade for a 1940s weekend. *Proving that sometimes even engines can dress up, 48773 is wearing mock air brakes. The War Department livery was cosmetic but historically authentic. 48773 was originally built for the War Department in 1940 to be used in France as WD307, but France was invaded by Germany before she could be exported.*

There are battle re-enactments, ARP Posts, hospital trains, singers performing 1940s acts, and even vintage fire-fighter displays by the NFS & AFS Vehicles Group which recreates the National Fire Service and the Auxiliary Fire Service from World War II.

The weekends remain popular, still attracting thousands of visitors twenty five years after they started.

Kidderminster station concourse with 1940s memorabilia. *In the background the sandbags of a temporary ARP shelter can just be seen.*

2000 – 2019: CRISES HIT THE LINE AND 'FAMILY FUN' BRANCHES OUT

Boiler Crisis

In May 2000 the board of the Holding Company was able to report that in 1999 the SVR had carried 228,010 passengers, the largest number since the line reopened and 7% higher than the year before. The railway also saw their highest turnover figure at £3.34 million with a retained profit of £55,532. (The small profit reflects the fact that the majority of money is ploughed back into running and improving the line.) At the time the report was written the carriage shed at Kidderminster (see p. 176) had just been completed and things looked positive.

Unfortunately, after the report was published a succession of the SVR's resident engines started to fail and had to be withdrawn due to boiler problems.

Through the summer, four of the ten engines in operation at the time failed, more than had ever happened before or indeed since. [11]

On 6th August two of the locomotives failed at the same time and, with two of the diesel fleet away for the Old Oak Common open day, the railway's working engines were down to only 7714, 1501, the diesel hydraulic D1013 and the five-car class 108 diesel multiple unit.

0-6-0 saddle tank 813 completed an overhaul in July 2007 and a month later was used to bolster the fleet during the boiler crisis. *The Hudswell Clarke engine built in 1900 pulled four coach trains for a week allowing the railway to retain steam on the timetable in the most critical period of the crisis.*

[11] None of the usual sources provide a complete list of the engines which failed. 1501 and 7325 definitely failed and, based on marked drops in their recorded mileages from surrounding years, it appears that 7802 and 45110 were also impacted. The author would be interested in hearing from anyone with concrete information.

Luckily, saddle tank 813 had just successfully run a single observation car on a private charter the day before, having returned to steam in July after a period of four years.

Despite the fact that she would not be able to haul a full seven coach train the locomotive department decided to put 813 to work on passenger services to protect as much of the timetable as possible, a job she did admirably.

Later the same day 1501's boiler failed as well, leaving only 7714 and the plucky 813 to keep things running for a week.

Despite the railway's best efforts to recover as quickly as possible a number of changes did have to be made through the year to accommodate the reduced fleet.

BR Standard 9F 2-10-0 92212 was hired in to run the Santa Specials in 2000 due to the boiler crisis in the summer. *Seen here at Hampton Loade on 11th February 2001.*

Although the Thomas event went ahead in October, the Autumn Steam Gala, the highlight of many an enthusiast's year, had to be cancelled for the first time in its history. And, as the full complement of resident engines were still not fully operational by December, 9F 2-10-0 92212 was drafted in to run the Santa Specials.

Once the crisis had passed it was discovered that there had been a number of causes for the failures, including the quality of the water used in the tanks. Through early 2001 reverse osmosis plants were installed at Bridgnorth and Kidderminster to address the water issues and reduce the likelihood of the crisis happening again.

Thankfully, due to the way it was handled (and of course the sterling efforts of 813 and the resident diesels!) the crisis did not impact passenger numbers too much. In the following year's report the Board was able to state that there had been only 785 fewer visitors than in the record setting 1999 (for comparison the number of passengers in 2018, the last published number at the time of writing, was 245,031).

The worst flooding in the line's history

The SVR has always been prone to landslips right from the very earliest days when it was being built, as we saw in Chapter One. The fact that it is located in a valley means that rainfall invariably flows towards the river, which in turn means that it has to cross the tracks, and there have been many times when flooding has affected the line. However in 2007 the railway experienced the worst flood damage in its one hundred year history.

On 19th June thunderstorms hit the area with a fortnight's worth of rain falling in just three quarters of an hour. Landing on already sodden ground it caused flash floods that tore away embankments, overflowed culverts.

By the next day it was clear that two thirds of the line, from Bridgnorth to Bewdley, was impassable to passenger services.

In June and July 2007 the railway saw the worst flood damage in its 100 year history. *Taken after the first floods on 19th June this picture shows the track bed having been swept away at Oldbury Viaduct looking down line to Bridgnorth.*

Repair work started almost immediately, with a public appeal for help being launched by 22nd, but a month later on 20th July there was another thunderstorm (the weather in between not having been much dryer) which caused further damage to areas already affected and added even more sites to the list.

The line between Kidderminster and Bewdley remained open throughout and, demonstrating both the determination of the railway and the level of support that was provided by other organisations during the crisis, it was even possible to stage a reduced Autumn Gala.

On the SVR itself of course it was only possible to use two locations for the gala: a small section of track at Bridgnorth which offered footplate rides on 42968, and the intact line between Kidderminster and Bewdley. However, due to the cooperation of Network Rail, services also ran from Bewdley along the main line to Birmingham Snow Hill.

Highley following the second storm *on 20th July. The track bed and embankment were swept downhill, damaging signalling equipment and the chalets visible below (thankfully nobody was injured).*

The repairs constituted the most extensive programme undertaken since the line was reopened, involving forty five separate locations that needed major work, including many that needed specialist outside contractors and significant work by the signalling department replacing equipment and cable. The stretch between Bridgnorth and Hampton Loade alone required more than two miles of signalling cable.

The line was fully reconnected on 10th March 2008, and passenger services resumed on 21st March.

The final cost of the repairs was estimated at £3.7 million, for which funding was provided by the railway's insurance, the public appeal and a series of grants from the European Regional Development Fund, Advantage West Midlands and the Heritage Lottery Fund. At the same time as repairing the line, work was also done to make the railway more able to withstand similar events in the future, for example a hundred new culverts were dug along the line to channel water through the track bed safely.

Borle Viaduct was the last of the affected areas to be repaired allowing the line to reopen on 21st March 2008. *Here A4 60007 'Sir Nigel Gresley' crosses the viaduct towards Highley in March 2009 a year after repair work was complete.*

Family fun branches out

The cost and requirements laid down by the company who license the *Day Out With Thomas* events meant that the last one was held in 2012, however, even before then the commercial success of these family events prompted the SVR's management to try to replicate it.

While Thomas obviously fits perfectly with the railway, several other children's favourites have therefore made an appearance on the line (often at the Engine House following its opening in 2007) including Peppa Pig and the Paw Patrol. Some have been less successful than others.

In 2000, at the height of the mania surrounding the Harry Potter books, 34027 Taw Valley (which was not then resident on the SVR) toured the country in red livery as 'Hogwarts Express' to promote the most recent novel, drawing crowds as predicted by the book's marketing folk.

In April of the following year the engine changed hands and came back to the SVR.

Seeing an opportunity to convert some young wizard fans to steam enthusiasts, the SVR ran it as the 'Hogwarts Express' in April, August and October 2001 but in October they were told by Warner Bros. (who own the intellectual property rights to the boy wizard) that the use was unsuitable and that the nameplates must be removed bringing the SVR's flirtation with Harry Potter experience to an end.

4-6-2 34027 'Taw Valley' was repainted in BR Green, *and had its 'Hogwarts Express' nameplate removed following Warner Bros. objection to its red Harry Potter livery in October 2001.*

In 2018 a new Wizard Express event was introduced, allowing families to enjoy the association they had come to make between heritage rail and magic. Children receive a wax sealed invitation to wizard school in exchange for their ticket and board a train where magicians perform tricks on the journey to wizarding school at Highley.

The event proved a real draw and returned in 2019.

CHAPTER TEN: Final Thoughts

When the group that became the SVR volunteers originally raised the idea of rescuing the line back in 1965 in the Coopers Arms, it is fair to say that they probably could not have imagined the work that would be involved in transforming their idea into a reality.

Nonetheless through a combination of passion and determination they got it off the ground and, against all odds, have kept it going successfully, fighting off what amounted to a fairly concerted takeover attempt, apathy and resistance from the local council, and even attempts by the weather to tear it all away!

Running the line as a commercial venture presented an altogether different set of hurdles, and occasionally required compromises that no doubt made the pure rail enthusiasts uncomfortable, however, it is clear that the volunteers and staff of the railway have risen to the challenge admirably.

In an article in *The Times* on 17th March 1972 John Slater, editor of the Railway Magazine and himself a huge proponent of preservation, repeated a criticism that was often levelled at heritage railways, namely that they were "selling rides not journeys".

It could be argued, I think, that over the years the Severn Valley Railway has gradually turned that criticism into a strength.

The railway has become an essential part of the local economy, bringing in tourism from far and wide, and offering a complete experience to their passengers who run the gamut from dedicated railway enthusiast, through people who might not be interested in the locomotives particularly but enjoy the 'living museum', to families who are only looking for a fun day out.

The spirit of love for the engines and the line that saw it reopened is still apparent on the faces of the volunteers and staff today, and must be a part of what brings so many of those different customers back again and again.

The goal for the railway going forward must be to maintain that spirit as the number of members and visitors who remember the days of steam on the mainline declines, but given the strong position it is starting from and its history so far, there is every reason to expect that it will be a challenge which will be met with enthusiasm and drive.

Three ages of rail at Bridgnorth: *Catch Me Who Can replica, 0-6-0 diesel electrical shunter D3586, and 7812 Erlestoke Manor.*
The Severn Valley's connection with railways has existed since the very earliest days of rail and looks set to survive for a long time yet to come.

ACKNOWLEDGEMENTS

Acknowledgements are traditionally where authors thank the people who've helped a book get written or published, but I want to start by thanking two men who had a much more fundamental role to play. Without them this book wouldn't exist at all.

Firstly of course Brian Lancashire, my father-in-law, to whom the book is dedicated.

Thank you for all of our fascinating discussions about steam and the travel stories you filled Verity's head with that have become so much a part of our life.

Obviously, I would have preferred it if we could have written this together like we'd planned, but have had to settle for doing it in your memory.

It wasn't the same, and I'm sure it is not as good.

Hopefully though it has enough that's interesting in it to prompt a few railway based discussions in other homes and we can keep it going.

And then Robert Williams (Uncle Bob), again for the conversations and imparting so much information (like the difference between geese and swans, and the fact that a pigeon special is a real thing!), but also a sincere thank you for all of the happy years of travel on the SVR.

You truly are a remarkably generous man and my family owe you more than we could ever hope to repay. If it hadn't been for your generosity and indulgence in allowing us to come with you to the SVR as often you did then there is no way we could have gone as regularly and a huge part of Connor, James and Bethany's childhoods would have been missing. It's even possible that my own interest would not have been sparked.

Now, with all of that gushing emotion out of the way, on to more normal acknowledgements:

My wife, Verity, and children, Connor, James and Beth have been their usual patient selves while I stole time away to write this. Thank you all, as always.

Thank you to the many contributors whose pictures have made it into the book, the overwhelming majority is work that people have chosen to make available for free online in exchange for simply being credited, they are all duly named in the dedicated Image Credits section, but a specific thank you is also called for. This book simply wouldn't have been possible without them, so thank you.

And in the same vein, I'd like to express my debt to the contributors to the Severn Valley Railway Wiki (svrwiki.com). To pull all of the information for the book together I have naturally used a large number of sources, but if any reader should want a single recommendation then the SVR Wiki would be it. It is a fantastic resource, crammed full of information on the most obscure pieces of the history of the line in preservation.

And finally, the tireless volunteers and staff who run the SVR. Writing this book has given me an even deeper appreciation for the work you do. Thank you for preserving such an essential part of our heritage and making it fun in the process.

Michael and Verity Lancashire
2019

REFERENCES / FURTHER READING

A wide variety of sources have been used in the research for this book, including published books from the nineteenth century onwards, privately published memoirs, historic newspapers and journals, archive news footage, modern websites and blogs, and even a couple of doctoral theses.

Rather than group them by type here, I have tried to organise them by topic so readers interested in any one area can find further reading which would most appeal.

SVR Specific Histories

Severn Valley Railway Memories, Christopher Magner, 1997

The Severn Valley Railway, John Marshall, 1989

Severn Valley Steam, Sir Gerald Nabarro, 1971

Handbook to the Severn Valley Railway Illustrative and Descriptive of Places along the Line from Worcester to Shrewsbury, J. Randall, 1863

The Severn Valley Railway, D.J. Smith, 1968

https://www.svrwiki.com

http://www.bewdleystation.co.uk/History.html

http://www.arleystation.org.uk/ashistory.html

General Railway Histories

A Guide to The Steam Railways of Great Britain, Rev. W. Awdry and Chris Cook (eds), revised by Roger Crombleholme, 1984

The Branch Lines of Worcestershire, Colin G. Maggs, 1994

Britain's Railways under Steam, J.B. Snell, 1965

The Great Western Railway: 150 Glorious Years, Patrick Whitehouse & David St. John Thomas, 1984

Passenger Services Timetable, British Railways, 1962

https://preservedbritishsteamlocomotives.com

Historical news

https://archive.cartoons.ac.uk/
https://www.britishnewspaperarchive.co.uk
https://www.macearchive.org
https://www.thetimes.co.uk/archive

Early history

Most information came from the SVR and General railway histories listed above, however, additional sources included:

Encyclopaedia Britannica, 15th edition
Sir Morton Peto, A Memorial Sketch by Henry Peto, 1893
The development of the railway network in Britain 1825-1911, Dan Bogart, Leigh Shaw-Taylor and Xuesheng You, paper from the University of Cambridge Group for the History of Population and Social Structure
Building Record: Kidderminster Railway Station, paper by Keith Beddoes and Bob Millward for Kidderminster & District Archaeological & Historical Society, 2014
https://www.gracesguide.co.uk
http://www.historyhome.co.uk/forpol/crimea/railway.htm
https://historicengland.org.uk
https://en.wikipedia.org

World War I & II

The Archaeology of World War II, paper by Malcolm Atkin of the Worcestershire Historic Environment and Archaeology Service
British Railways and the War, Fred A. McKenzie, 1917
Operation Pied Piper: The Wartime Evacuation of Schoolchildren from London and Berlin 1938-46, Niko Gärtner, PhD Thesis University of London, 2011
https://media.nationalarchives.gov.uk/index.php/railways-and-the-mobilisation-for-war-in-1914/
https://www.iwm.org.uk/history/transport-and-supply-during-the-first-world-war
https://millsarchive.org/explore/features-and-articles/entry/197975/mills-at-war=/14708
https://www.networkrail.co.uk/stories/wwi-and-the-railway/
http://www.rail.co.uk/rail-news/2014/wartime-railways-the-first-months/
https://www.railwaymuseum.org.uk/objects-and-stories/ambulance-trains-bringing-first-world-war-home
https://www.warwickshirerailways.com/
Information about Private Brady was uncovered on www.ancestry.co.uk and historical newspapers (see above)

Modern era

Exploits of a Politician, Sir Gerald Nabarro, 1973

Fifty years of motorways: how did we get there and where do we go to next?, paper by Professor David Bayliss OBE of Halcrow Consulting and Imperial College London, 2009

https://alveleymininghéritage.org

http://www.alveleyhistoricalsociety.com/mining.html

http://www.bbc.co.uk/herefordandworcester

https://www.dailyecho.co.uk/news/3703027.nabarro-nabbed-for-danger-driving/

https://www.expressandstar.com

https://www.facebook.com/svrofficial

https://forum.svr-online.org.uk

http://www.kfriends.org.uk/magazine/engineering%20report01.htm

https://www.kidderminstershuttle.co.uk

http://www.highleystation.co.uk

https://www.metoffice.gov.uk/binaries/content/assets/metofficegovuk/pdf/weather/learn-about/uk-past-events/interesting/2007/heavy-rainfall_flooding---june-2007---met-office.pdf

http://www.severntales.co.uk/worcestershirebridges.html

https://www.sixbellsjunction.co.uk

https://www.svr.co.uk

https://www.svrlive.com

http://svr-vlo.org.uk

https://www.svrtrust.org.uk

https://www.thetimes.co.uk/archive/article/1973-06-29/1/10.html

https://www.worcesternews.co.uk

Monetary values

All conversions into today's values have used the MeasuringWorth website. The calculator provides a range of answers including straight RPI inflation and purchasing power, for consistency I have used the lowest value in each case.

https://www.measuringworth.com/calculators/ukcompare

Legal document (e.g. company accounts, Light Railway Orders, etc.)

http://www.legislation.gov.uk/all?title=severn%20valley

https://api.parliament.uk/historic-hansard/commons/index.html

https://www.railwaysarchive.co.uk

https://beta.companieshouse.gov.uk/search?q=severn+valley+railway

IMAGE CREDITS

The majority of the pictures used in this book were released under Creative Commons. The creative commons license can be found at https://creativecommons.org/licenses/by-sa/2.0/

All of the images have been attributed where appropriate, but if you believe an attribution is incorrect then please tell me via an email to contactus@sannox.org.uk (if some of the copyright holder's names look a little odd then remember that in many cases they are internet handles!)

The information for each picture is organised as follows: *Page number (in bold), Subject, Source or Link, Copyright holder, Any relevant notes.* Entries are separated by a semi-colon.

Front Cover, https://flic.kr/p/ip1M12, Neil Scott; Back Cover, https://www.geograph.org.uk/photo/4505405, Keith Wilkinson; Frontispiece, https://en.wikipedia.org/wiki/File:5164_Victoria_Bridge_Severn_Valley_Railway.jpg, Duncan Harris; **8**, Catch Me Who Can advertisement, https://en.wikipedia.org/wiki/Catch_Me_Who_Can#/media/File:Catch_me_who_can.jpg, Copyright holder unknown; **8**, Richard Trevithick, https://commons.wikimedia.org/wiki/File:Richard_Trevithick_portrait.jpg, John Linnell in the Science Museum collection, Public Domain; **9**, Bradshaw railway map of GB 1853, https://doi.org/10.3931/e-rara-38938, Copyright holder unknown; **10**, Sir Samuel Morton Peto Bt, taken from *Sir Morton Peto, A Memorial Sketch* by Henry Peto Copyright holder unknown; **11**, Houses of Parliament, https://picryl.com/media/houses-of-parliament-london-england-1, Photochrom Print Collection; **12**, Commencement of the railway works at Balaclava from Illustrated London News, 10 March 1855, https://commons.wikimedia.org/wiki/File:Crimean_Railway_4.jpg, Copyright holder unknown; **16**, LCDR, https://flic.kr/p/5xbsxh, Andrew

Bowden; **17**, Sir Samuel Morton Peto Bt in later life, taken from *Sir Morton Peto, A Memorial Sketch* by Henry Peto Copyright holder unknown;**18**, John Fowler, https://wellcomecollection.org/works/bzqaujxx?wellcomeImagesUrl=/indexplus/image/V0026397.html#licenseInformation, Wellcome Collection Photo by Lock & Whitfield; **18**, Metropolitan Line advert, https://commons.wikimedia.org/wiki/File:Metropolitan_Underground_Railway_stations.jpg, Copyright holder unknown, Public Domain; **19**, Forth bridge, Plate from W. Westhofen, The Forth Bridge, https://archive.org/details/forthbridgethewe00westrich/page/n4, Copyright holder unknown; **20**, 42968 showing bends in line near Highley, https://flic.kr/p/djHqmX, Tony Hisgett; **21**, Bradshaw's Railway Handbook 1863, author's collection; **23**, GWR coat of arms, https://commons.wikimedia.org/wiki/File:Coat_of_arms_of_the_Great_Western_Railway.png, Sandy JJ Gould; **25**, 6695 pulls a goods train, https://flic.kr/p/7HG2Au, Duncan Harris; **26**, 2857 at Bewdley south signal box, https://flic.kr/p/ip1M12, Neil Scott; **28**, 3205, https://flic.kr/p/JTZeye, foundin_a_attic; **29**, 4566 at Kidderminster, https://flic.kr/p/e5zdEL, Tony Hisgett; **30**, 5164, https://flic.kr/p/7HG43Q, Duncan Harris; **31**, Early railcar, https://commons.wikimedia.org/wiki/File:Enfield_Bridges_Adams_steam_railmotor.jpg, Copyright holder unknown, Public Domain; **32**, GWR railcar, https://commons.wikimedia.org/wiki/File:GWR_Steam_Railmotor_No_93_At_the_Didcot_Railway_Centre,_cropped.jpg, Copyright holder unknown, Public Domain; **33**, 2047 Warwickshire, https://flic.kr/p/dZq9Ry, Hugh Llewelyn; **33**, 7714 at the water crane, https://geograph.org.uk/p/6092871, Chris Allen; **37**, Birmingham Gazette Declaration of WWI, https://blog.britishnewspaperarchive.co.uk/2014/07/30/historic-headlines-great-britain-joins-world-war-one-on-4-august-1914/; **37**, Sir Herbert Walker from Railway Magazine July 1915, https://commons.wikimedia.org/wiki/File:Sir_Herbert_Ashcombe_Walker.jpg, Copyright holder unknown, Public Domain; **38**, REC minutes, https://www.railwaysarchive.co.uk, Ministry of Transport, Public Domain, Crown Copyright expired; **39**, Boer War ambulance train, https://wellcomecollection.org/works/ag79wnwq, Wellcome Collection Photo by Major Henry Brazier Creagh; **40**, Ambulance train, https://digital.nls.uk/first-world-war-official-photographs/archive/74548610#?c=0&m=0&s=0&cv=29&xywh=0%2C-73%2C2499%2C2009, David McLellan; **40**, Interior of ambulance train, https://digital.nls.uk/first-world-war-official-photographs/archive/74548610#?c=0&m=0&s=0&cv=27&xywh=-963%2C-186%2C4425%2C3702, David McLellan; **42**, GWR Roll of honour, https://flic.kr/p/aDU27c, (Mick Baker)rooster; **43**, Elan Valley Aqueduct, https://geograph.org.uk/p/152047, Mr M Evison; **46**, Sir Eric Geddes, https://commons.wikimedia.org/wiki/File:Sir_Eric_Campbell-Geddes_in_1917.jpg, George Grantham Bain Collection; **47**, Railways Act 1921, https://www.railwaysarchive.co.uk/documents/HMG_Act_Reg1921.pdf, Act of Parliament; **48**, Eardington, https://www.geograph.org.uk/more.php?id=3752733, Nigel Thompson; **49**, Sir Ralph Wedgwood Bt, https://www.gracesguide.co.uk/Ralph_Wedgwood, Copyright holder unknown; **50**, "In war and peace we serve" poster, https://www.railwaysarchive.co.uk/docsummary.php?docID=2091, Ministry of Transport, Public Domain, Crown Copyright expired; **50**, "Another mechanised army" poster, https://www.railwaysarchive.co.uk/docsummary.php?docID=2090, Ministry of Transport, Public Domain, Crown Copyright expired; **51**, Evacuation from Waterloo, https://commons.wikimedia.org/wiki/File:A_scene_at_a_London_rail way_station_showing_troops_arriving_while_kiddies_who_are_being_evacuated_from_London_leave_for_the_reception_area_HD-SN-99-02672.jpg, Copyright holder unknown; **51**, "Leave this to us" poster, https://commons.wikimedia.org/wiki/File:Leave_this_to_Us_Sonny_Art. IWMPST13854.jpg, Copyright holder unknown; **52**, Down Street exterior, https://flic.kr/p/7MTjKp, Ewan Munro; **53**, Down Street REC HQ plan, image from information boards; **54**, REC committee room, https://www.subbrit.org.uk/sites/down-street-station/, Copyright holder unknown; **55**, Spitfires and airmen, https://commons.wikimedia.org/wiki/File:123_Squadron_RAF_Spitfires_and_pilots_Castletown_WWII_IWM_CH_4180.jpg, Copyright holder unknown, Public Domain, Crown Copyright expired; **56**, General Patton, https://pixabay.com/photos/general-george-s-patton-jr-military-402041/, Public Domain; **60**, Transport Act 1947, http://www.legislation.gov.uk/ukpga/Geo6/10-11/49/contents/enacted, Act of Parliament; **61**, BR logo, https://commons.wikimedia.org/wiki/File:British_Rail_Flying_Sausage_log o.png, cnbrb; **61**, 1949 BR emblem, https://commons.wikimedia.org/wiki/File:Drewry_No.2589,_BR_No.1 2589_(12748818505).jpg, Hugh Llewelyn; **62**, Modernisation plan, https://www.railwaysarchive.co.uk/docsummary.php?docID=23, British Transport Commission, Public Domain; **64**, Beeching Report Part 1, https://www.railwaysarchive.co.uk/docsummary.php?docID=13, British Railways Board, Public Domain; **65**, Beeching Report map of lines to be closed, https://www.railwaysarchive.co.uk/docsummary.php?docID=35, British Railways Board, Public Domain; **68**, Shrewsbury station (3204 at platform), https://www.geograph.org.uk/photo/2540220, Ben Brooksbank; **69**, SVR extract from Beeching Report, https://www.railwaysarchive.co.uk/docsummary.php?docID=13, British Railways Board, Public Domain; **70** DMU at Coalport, https://www.geograph.org.uk/photo/335267, Dr Neil Clifton; **70** Special last train, https://www.svrwiki.com/The_Severn_Valley_Railway_under_GWR/BR_ownership#/media/File:Bridgnorth-Last-Train-1963-09-08.jpg, Sellick Collection; **72**, Stepney on Bluebell, https://commons.wikimedia.org/wiki/File:55StepneyHorstedKeynes.JP G, bluebellnutter; **74-77**, Working Parties on the line, http://www.bbc.co.uk/herefordandworcester/content/image_galleries/sever n_valley_railway_archive_gallery.shtml?1, Copyright holder unknown; **78**, 3442, https://flic.kr/p/dztkPa, Hugh Llewelyn; **79**, 4555, https://en.m.wikipedia.org/wiki/File:Witham_(Somerset)_Station,_with_a_train_for_Yatton_geograph-2545301-by-Ben-Brooksbank.jpg, Ben Brooksbank; **83**, 3205 at Hampton Loade, https://www.geograph.org.uk/photo/4299997,

Roger Cornfoot; **84**, Daniel's mill, https://www.geograph.org.uk/photo/1491907, K A; **84**, 46443, https://www.geograph.org.uk/photo/2781621, Ben Brooksbank; **85**, 46443, https://flic.kr/p/dSGwUe, Hugh Llewelyn; **85**, Railway man's arms, https://www.geograph.org.uk/photo/4703249, Roger Kidd; **86**, W22W, https://flic.kr/p/dSwfxR, Hugh Llewelyn; **87**, 2047, https://flic.kr/p/dZq995, Hugh Llewelyn; **88**, Guarantee Co incorporation, https://beta.companieshouse.gov.uk/company/00906842/filing-history, Public Domain; **88**, 813, https://www.geograph.org.uk/photo/5732793, Philip Pankhurst; **89**, 47383, https://commons.wikimedia.org/wiki/File:47383_at_Highley.jpg, David Ingham; **90**, 43106, https://flic.kr/p/dZHJz1, Hugh Llewelyn; **92**, 3205 at Bridgnorth, https://www.geograph.org.uk/photo/2932617, Ben Brooksbank; **93**, 43106 on opening day, https://www.geograph.org.uk/photo/2781931, Ben Brooksbank; **93**, 46443 on opening day, https://www.geograph.org.uk/photo/2781959, Ben Brooksbank; **96**, 1501, https://flic.kr/p/o6GtRL, Juan Enrique Gilardi; **97**, Nabarro signature, Author's collection, Verity Lancashire; **98**, 45110 April 1971, https://flic.kr/p/dZJwTG, Hugh Llewelyn; **99**, 70000, https://flic.kr/p/2adR1BE, Tony Hisgett; **102**, 600, https://flic.kr/p/D5t93b,Hugh Llewelyn; **103**, Holding Co incorporation, https://beta.companieshouse.gov.uk/company/01046274/filing-history, Public Domain; **104**, 61994, https://flic.kr/p/cWtev3, Andrew; **105**, Nabarro cartoon, https://archive.cartoons.ac.uk/record.aspx?src=CalmView.Catalog&id=23318, Barry Lewis; **108**, 4930, https://flic.kr/p/bzbhDB, Barry Lewis; **110**, 7819, https://flic.kr/p/bzpFUt, Barry Lewis; **111**, 78019, https://flic.kr/p/7sLLac, Duncan Harris; **115**, 75069, https://flic.kr/p/dJi3fm, 70023venus2009; **116**, 42968, https://flic.kr/p/7SgpM2, Roland Turner; **117**, 43106 at Bewdley, https://www.geograph.org.uk/photo/2865564, Gareth James; **118**, 5764 on Victoria Bridge, https://flic.kr/p/7A1Keu, Les Chatfield; **119**, 46521 at Barry, https://flic.kr/p/dARTbs, Hugh Llewelyn; **119**, 46521 restored, https://www.geograph.org.uk/photo/767376, John Firth; **120**, 6960, https://www.geograph.org.uk/photo/2743912, Rob Newman; **120**, 7812, https://flic.kr/p/vNqV7R, jess_k_kent1; **120**, 7802, https://flic.kr/p/mLs8zP, Hazel Nicholson; **121**, Bridgnorth by-pass, https://geograph.org.uk/p/254197, David Stowell; **122**, 80079, https://www.geograph.org.uk/photo/2480214, Rob Newman; **123**, 600, https://flic.kr/p/eYQHwx,Hugh Llewelyn; **124**, Kidderminster Goods Yard Feb 1984, https://www.geograph.org.uk/photo/248565, David Stowell; **125**, Kidderminster Station construction May 1984, https://www.geograph.org.uk/photo/248539, David Stowell; **126**, Light Railway Order, https://www.legislation.gov.uk/uksi/1984/1202/pdfs/uksi_19841202_en.pdf, Public Domain; **126**, 4930 Inaugural Train from Kidderminster 30/07/84, https://www.geograph.org.uk/photo/248546, David Stowell; **127**, Leander at newly opened Kidderminster 07/10/84, https://www.geograph.org.uk/photo/1111830, P L Chadwick; **131**, Bridgnorth without loco shed 02/09/1972, https://www.geograph.org.uk/photo/3269207, Ben Brooksbank; **131**, Bridgnorth with loco shed but no boiler shop

08/04/1984, https://www.geograph.org.uk/photo/1926329, John Baker; **132**, The Engine Shed, https://www.geograph.org.uk/photo/5732745, Philip Pankhurst; **133**, Bridgnorth with boiler shop 26/04/2015, https://www.geograph.org.uk/photo/4449611, Stephen McKay; **134**, Inside the Boiler shop, https://www.geograph.org.uk/photo/5731641, Philip Pankhurst; **135**, Bridgnorth refreshment room, https://www.svrwiki.com/Bridgnorth#/media/File:Bridgnorth_New_Building_20181229.jpg, Robin; **136**, 3717 at Crossing Cottage, https://www.geograph.org.uk/photo/225002, Neil Kennedy; **137**, Eardington station 1994, https://www.geograph.org.uk/photo/4098530, John Winder; **138**, Eardington station restored, https://www.geograph.org.uk/photo/5796882, Philip Pankhurst; **140**, Hampton Loade station, https://www.geograph.org.uk/photo/3752719, Nigel Thompson; **141**, Paddock Railway, https://www.geograph.org.uk/photo/2504951, Richard Law; **141**, 7802 at Hampton Loade, https://www.geograph.org.uk/photo/206393, Neil Kennedy; **143 & 144**, Historical pictures of Alveley colliery and bridge, http://www.alveleyhistoricalsociety.com/mining.html, Copyright holder unknown; **144**, 5764 at Country Park Halt, https://www.geograph.org.uk/photo/5558, David Stowell; **145**, Highley station house, https://www.geograph.org.uk/photo/2183926, Nick Macneill; **146**, 34027 and Highley signal box, https://www.geograph.org.uk/photo/4646718, David P Howard; **146**, 1450 on no. 2 siding at Highley, https://www.geograph.org.uk/photo/4395686, Chris Allen; **147**, Engine House from line side, https://www.geograph.org.uk/photo/3982583, David P Howard; **148**, Engine house, https://flic.kr/p/oaHthy, Juan Enrique Gilardi; **149**, 5164 passing under Highley footbridge, https://www.geograph.org.uk/photo/2746263, Rob Newman; **150**, Upper Arley, https://www.geograph.org.uk/photo/1492087, K A; **151**, Arley footbridge, https://www.geograph.org.uk/photo/3243733, Jaggery; **152**, Ferry in 1924 , https://flic.kr/p/nHcGth, Bessie Hayward; **153**, Arley platform, https://www.geograph.org.uk/photo/2501429, Richard Law; **155**, Victoria Bridge through trees, https://www.geograph.org.uk/photo/618665, Ant Collins; **156**, Victoria Bridge detail, https://commons.wikimedia.org/wiki/File:7802_Bradley_Manor_01.jpg, Mightyshrew; **157**, 3205 at Northwood Halt, https://www.geograph.org.uk/photo/2187528, K A; **158**, Level crossing at Northwood, https://www.geograph.org.uk/photo/5533, David Stowell; **161**, Bewdley station, https://flic.kr/p/THfPDP, Hugh Llewelyn; **162**, Wribbenhall viaduct, https://www.geograph.org.uk/photo/3169126, Gareth James; **163**, 60019 and Bewdley island platform, https://www.geograph.org.uk/photo/2865569, Gareth James; **164**, Bewdley token exchange, https://flic.kr/p/2evqmFy, Jeremy Segrott; **165**, 4566 at Bewdley, https://flic.kr/p/aaA2fj, 70023venus2009; **166**, 34053 on loco inspection pit at Bewdley, https://flic.kr/p/BPTwX5, Andrew; **167**, 2857 with goods train, https://www.geograph.org.uk/photo/6088951, Fabian Musto; **169** Kidderminster BR Station building 1963, https://www.geograph.org.uk/photo/3387931, Ben Brooksbank; **169**, Kidderminster BR Station building 2006, https://www.geograph.org.uk/photo/118927, David Stowell; **170**, Kidderminster Town Station, https://www.geograph.org.uk/photo/1829879, P L

Chadwick; **171**, 1501 and 46443 in Kidderminster Town, https://www.geograph.org.uk/photo/2320383, Rob Newman; **172**, Concourse roof, https://flic.kr/p/XYwSX6, Tony Hisgett; **173**, 45690 Leander passing Kidderminster signal box, https://flic.kr/p/8DSzEW, Tony Hisgett; **174**, Kidderminster signal levers, https://www.svrwiki.com/File:Kidderminster_signal_box_interior.jpg, Graham Phillips; **175**, 7812 in front of carriage works, https://www.geograph.org.uk/photo/2467736, Philip Pankhurst; **176**, Dynamo rig at carriage works, https://www.geograph.org.uk/photo/6142181, Chris Allen; **177**, Carriage shed, https://www.geograph.org.uk/photo/1557177, John M; **178**, Inside carriage shed, https://www.geograph.org.uk/photo/5464556, Philip Pankhurst; **180**, Sherlock Holmes 2011 movie poster, http://www.impawards.com/2011/sherlock_holmes_a_game_of_shadows_ver18.html, Warner Bros. Pictures, used under Fair Dealing; **181**, 60103 Flying Scotsman at Eardington, https://flic.kr/p/MyLc4e, Brian Taylor; **182**, 6024, https://www.geograph.org.uk/photo/840393, Neil Kennedy; **184**, 45110 at Arley, Author's collection, Verity Lancashire; **187**, 686 shortly after arrival, https://flic.kr/p/dZqau5, Hugh Llewelyn; **187**, 686 as Thomas, Author's collection, Verity Lancashire; **187**, 686 at the Engine House September 2011, https://flic.kr/p/avtE8Z, Hugh Llewelyn; **188**, 48773 as WD307 at Hampton Loade, https://www.geograph.org.uk/photo/2310470, Rob Newman; **189**, ARP Post Kidderminster, https://www.geograph.org.uk/photo/2499789, Richard Law; **190**, 813, https://www.geograph.org.uk/photo/5616763, Mary and Angus Hogg; **191**, 92212 at Hampton Loade, https://www.geograph.org.uk/photo/2746270, Rob Newman; **193**, Flood damage at Oldbury Viaduct, https://www.geograph.org.uk/photo/552051, David Stowell; **194**, Highley flood damage, https://www.geograph.org.uk/photo/552441, David Stowell; **195**, https://www.geograph.org.uk/photo/2710907, K A; **196**, 34027 outside Highley, https://www.geograph.org.uk/photo/5137076, Gareth James; **198**, Three ages of rail, https://flic.kr/p/5RC1Ci, Andrew Northall

INDEX OF SVR STEAM LOCOMOTIVES

Locomotives resident on the SVR at the time of writing (January 2020) have been marked in two ways: engines in service are highlighted in bold, those undergoing restoration or overhaul are marked in italics.

Where an image of the loco has been included in the book, page numbers for the images are given. A full index of references to engines in the text can be found at the back of the index.

Engine	Pic Page	Whyte	Class or type	Original owner	Location
193 Shropshire	n/a	0-6-0 ST	Austerity	WD	Ribble Steam Railway
417 Invicta	n/a	0-4-0 ST	n/a	Chatham Dockyard	Chatham Historic Dockyard Railway
600 Gordon	102 123	2-10-0	WD "Austerity"	Longmoor Military Railway	SVR
686 The Lady Armaghdale	187	0-6-0 T	n/a	Manchester Ship Canal Railway	SVR

813	**190**	**0-6-0 ST**	**n/a**	**Port Talbot Railway**	**SVR**
1000	n/a	4-4-0	Johnson 4P Midland Compound	MR	Barrow Hill
1450	**146**	**0-4-2 T**	**1400**	**GWR**	**SVR**
1501	**96, 171**	**0-6-0 PT**	**1500**	**BR**	**SVR**
1738	n/a	0-4-0 ST	n/a	CEGB	Titley Junction Station
2047 Warwickshire	*33, 84, 87*	*0-6-0 ST*	*n/a*	*Rugby Portland Cement*	*SVR*
2857	**26, 167**	**2-8-0**	**2800**	**GWR**	**SVR**
3020 Cornwall	n/a	2-2-2	n/a	LNWR	NRM Shildon
3205	28, 83, 92, 157	0-6-0	2251	GWR	South Devon Railway
3717 City of Truro	136	4-4-0	3700 City	GWR	NRM York
4141	n/a	2-6-2 T	5101 Large Prairie	GWR	Epping Ongar Railway
4150	*n/a*	*2-6-2 T*	*5101 Class Large Prairie*	*GWR*	*SVR*
4566	29, 165	2-6-2 T	4500 Class Small Prairie	GWR	SVR
4930 Hagley Hall	*108 126*	*4-6-0*	*4900 Class Hall*	*GWR*	*SVR*
5164	30, 149	2-6-2 T	5101 Class Large Prairie	GWR	SVR
5474 Carnarvon	n/a	0-6-0 ST	n/a	Stewarts & Lloyds	South Devon Railway
6634	n/a	0-6-2 T	5600 Class	GWR	Peak Rail
6960 Raveningham Hall	120	4-6-0	6959 Modified Hall	GWR	West Somerset Railway

7325	n/a	2-6-0	4300 Class Mogul	GWR	SVR
7714	**33**	**0-6-0 PT**	**5700 Class Pannier**	**GWR**	**SVR**
7802 Bradley Manor	**120 141**	**4-6-0**	**7800 Manor**	**GWR**	**SVR**
7812 Erlestoke Manor	*120 175 198*	*4-6-0*	*7800 Manor*	*GWR*	*SVR*
7819 Hinton Manor	110	4-6-0	7800 Manor	GWR	SVR
34027 Taw Valley	**146 196**	**4-6-2**	**Bulleid Battle of Britain Class**	**SR**	**SVR**
34053 Sir Keith Park	162 166	4-6-2	Bulleid Battle of Britain Class	SR	Swanage Railway
42968	*20, 116 134*	*2-6-0*	*Stanier Mogul*	*LMS*	*SVR*
43106	**90, 93, 117, 119, 132**	**2-6-0**	**Ivatt Class 4**	**BR**	**SVR**
45000	n/a	4-6-0	Stanier 5MT Black 5	LMS	NRM Shildon
45110	98, 184	4-6-0	5MT Black Five	LMS	SVR
45690 Leander	127 173	4-6-0	Stanier 6P Jubilee	LMS	WCR Carnforth
45699 Galatea	n/a	4-6-0	Stanier 6P Jubilee	LMS	WCR Carnforth
46443	84, 85, 93, 171	2-6-0	Ivatt Class 2 "Mickey Mouse"	BR	SVR
46521	119	2-6-0	Ivatt Class 2 Mickey Mouse	LMS	Great Central Railway
47383	89	0-6-0 T	Fowler 3F "Jinty"	LMS	SVR
48773	188	2-8-0	Stanier 8F	LMS	SVR
60009 Union of South Africa	n/a	4-6-2	Gresley A4 P acific	LNER	Scotland

61994 The Great Marquess	78, 104	2-6-0	Gresley K4 West Highland	LNER	Bo'Ness
70000 Britannia	99	4-6-2	Riddles Class 7	BR	Icons of Steam
75069	**115**	**4-6-0**	**Riddles 4M T**	**BR**	**SVR**
78019	111	2-6-0	Riddles 2MT	BR	Great Central Railway
80079	122	2-6-4 T	Riddles 4MT	BR	SVR
3612	n/a	0-6-0 PT	5700 Pannier	GWR	Scrapped
5786	n/a	0-6-0 PT	5700 Pannier	GWR	South Devon Railway
64 305	n/a	2-6-2 T	DRG Class 64	DRG	Nene Valley Railway
L95 (5764)	118 144	0-6-0 PT	5700 Class Pannier	GWR	SVR

INDEX

(Page numbers in bold indicate images)

Topics

L

M

N

O

P

R

S

T

ABOUT THE AUTHOR

Michael Lancashire has previously written a fiction series about a criminal mastermind known as the Architect, and in 2017 released the first in a series of spy books aimed at young adults.
Severn Valley Railway: An Illustrated History is his first non-fiction book.

He lives with his wife and three children, splitting their time between the Midlands and a Scottish island hideaway.

The books featuring the Architect are:
The Voynich Deception
Kernel Panic
The Keycode: An Architect Short Story
The Fourth Target
Almost Out
Trapped

He has also written a British political thriller:
Heritage: A short story

And, for young adults:
Jack Hawke in Killer Game

Links to all of the above at your favourite store are available on his website sannox.org.uk

Printed in Great Britain
by Amazon

71486742R00132